MHC Volume 2

The Practical Approach Series

SERIES EDITORS

D. RICKWOOD
Department of Biology, University of Essex, Wivenhoe Park, Colchester, Essex CO4 3SQ, UK

B. D. HAMES
Department of Biochemistry and Molecular Biology University of Leeds, Leeds LS2 9JT, UK

★ indicates new and forthcoming titles

Affinity Chromatography
★ Affinity Separations
Anaerobic Microbiology
Animal Cell Culture (2nd edition)
Animal Virus Pathogenesis
Antibodies I and II
★ Antibody Engineering
★ Antisense Technology
★ Applied Microbial Physiology
Basic Cell Culture
Behavioural Neuroscience
Bioenergetics
Biological Data Analysis
Biomechanics – Materials
Biomechanics – Structures and Systems
Biosensors
Carbohydrate Analysis (2nd edition)
Cell-Cell Interactions
The Cell Cycle
Cell Growth and Apoptosis
Cellular Calcium
Cellular Interactions in Development
Cellular Neurobiology
Clinical Immunology
★ Complement
Crystallization of Nucleic Acids and Proteins
Cytokines (2nd edition)
The Cytoskeleton
Diagnostic Molecular Pathology I and II
★ DNA and Protein Sequence Analysis
DNA Cloning 1: Core Techniques (2nd edition)
DNA Cloning 2: Expression Systems (2nd edition)
DNA Cloning 3: Complex Genomes (2nd edition)
DNA Cloning 4: Mammalian Systems (2nd edition)
Electron Microscopy in Biology

Electron Microscopy in Molecular Biology

Electrophysiology

Enzyme Assays

★ Epithelial Cell Culture

Essential Developmental Biology

Essential Molecular Biology I and II

Experimental Neuroanatomy

Extracellular Matrix

Flow Cytometry (2nd edition)

★ Free Radicals

Gas Chromatography

Gel Electrophoresis of Nucleic Acids (2nd edition)

Gel Electrophoresis of Proteins (2nd edition)

Gene Probes 1 and 2

Gene Targeting

Gene Transcription

★ Genome Mapping

Glycobiology

Growth Factors

Haemopoiesis

Histocompatibility Testing

HIV Volumes 1 and 2

Human Cytogenetics I and II (2nd edition)

Human Genetic Disease Analysis

★ Immunochemistry 1

★ Immunochemistry 2

Immunocytochemistry

In Situ Hybridization

Iodinated Density Gradient Media

Ion Channels

Lipid Modification of Proteins

Lipoprotein Analysis

Liposomes

Mammalian Cell Biotechnology

Medical Parasitology

Medical Virology

★ MHC Volumes 1 and 2

Molecular Genetic Analysis of Populations

Molecular Genetics of Yeast

Molecular Imaging in Neuroscience

Molecular Neurobiology

Molecular Plant Pathology I and II

Molecular Virology

Monitoring Neuronal Activity

Mutagenicity Testing

★ Neural Cell Culture

Neural Transplantation

★ Neurochemistry (2nd edition)

Neuronal Cell Lines

NMR of Biological Macromolecules

Non-isotopic Methods in Molecular Biology

Nucleic Acid Hybridisation

Oligonucleotides and Analogues

Oligonucleotide Synthesis

PCR 1

PCR 2

★ PCR In Situ Hybridization

Peptide Antigens

Photosynthesis: Energy Transduction
Plant Cell Biology
Plant Cell Culture (2nd edition)
Plant Molecular Biology
Plasmids (2nd edition)
★ Platelets
Postimplantaton Mammalian Embryos
Preparative Centrifugation
Protein Blotting
Protein Engineering
★ Protein Function (2nd edition)
Protein Phosphorylation
Protein Purification Applications
Protein Purification Methods
Protein Sequencing
★ Protein Structure (2nd edition)
★ Protein Structure Prediction
Protein Targeting
Proteolytic Enzymes
Pulsed Field Gel Electrophoresis
RNA Processing I and II
★ Signalling by Inositides
★ Subcellular Fractionation
Signal Transduction
Transcription Factors
Tumour Immunobiology

MHC
Volume 2
A Practical Approach

Edited by

N. FERNANDEZ
Department of Biological and Chemical Sciences, University of Essex

and

G. BUTCHER
Babraham Institute, Babraham, Cambridge

Oxford New York Tokyo

Oxford University Press, Great Clarendon Street, Oxford OX2 6DP

Oxford New York
Athens Auckland Bangkok Bogota Bombay Buenos Aires
Calcutta Cape Town Dar es Salaam Delhi Florence Hong Kong
Istanbul Karachi Kuala Lumpur Madras Madrid Melbourne
Mexico City Nairobi Paris Singapore Taipei Tokyo Toronto Warsaw

and associated companies in
Berlin Ibadan

Oxford is a trade mark of Oxford University Press

Published in the United States
by Oxford University Press Inc., New York

A catalogue record for this book is available from the British Library

Library of Congress Cataloging in Publication Data
(Data available)

ISBN 0 19 963556 0 (Hbk)
ISBN 0 19 963555 2 (Pbk)

Two volume set ISBN 0 19 963558 7 (Hbk)
ISBN 0 19 963557 9 (Pbk)

Typeset by Footnote Graphics, Warminster, Wilts
Printed in Great Britain by Information Press Ltd, Eynsham, Oxon.

Foreword

by Jack L. Strominger

In the early 1970s when sustained efforts to characterize transplantation antigens began, the number of investigators who were active in this field could be counted on the fingers of both hands, or at most fingers and toes. This book will be a wonder to this group of scientists. Over the next 25–30 years the field developed enormously. The molecules were renamed major histocompatibility antigens. They are encoded in a single genetic region on chromosome 6 in man or chromosome 17 in the mouse, called the major histocompatibility complex (MHC), that provides the major barrier to tissue transplantation. Two classes of major histocompatibility proteins were identified and were later shown to be critical elements in the initiation of an immune response (organ rejection being a by-product of this function); a third class includes components of the complement system. Thus, this region is central to the development of immune responses in all higher vertebrates and at least some lower vertebrates. Class I and Class II MHC proteins were defined, isolated, and crystallized and their 3-dimensional structures were determined, but that was only the beginning. When it was found that these molecules present peptides derived from self and foreign proteins to the immune system, then many additional facets began to be investigated. How are antigenic proteins degraded? How are the peptides processed and loaded on to the class I and class II MHC proteins? Where does this loading occur and how do the peptides get there? What kinds of auxiliary molecules are required on antigen presenting cells? Do different antigen presenting cells have different properties? How does one obtain lines and clonal populations of the effector cells (T cells and Natural Killer cells) that respond to alterations in MHC/peptide complexes? Do T cells and Natural Killer cells interact with MHC proteins in the same manner? What are the bases for alloreactivity (that leads to graft rejection) and for the severe biological effect of some bacterial exotoxins (superantigens that interact with Class II MHC proteins)?

In addition to containing many practical details concerning these areas of investigation, this book contains a wealth of general information. It will provide a very useful manual for the many scientists now working in the field and for those intending to enter it, as well as glimpses of areas that remain to be explored.

Contents

Contributors

ANDERS BRUNMARK
R. W. Johnson Pharmaceutical Research Institute, San Diego, CA 92121, USA.

RICHARD J. CHERRY
University of Essex, Wivenhoe Park, Colchester CO4 3SQ, UK.

SHIRLEY A. ELLIS
Institute for Animal Health, Compton, Nr Newbury, Berkshire RG20 7NN, UK.

NELSON FERNANDEZ
University of Essex, Department of Biological and Chemical Sciences, Wivenhoe Park, Colchester CO4 3SQ, UK.

GIOVANNI BATTISTA FERRARA
Immunogenetic Laboratory, National Institute for Cancer Research, Largo Rosanna Benzi, 10, 16132 Genova, Italy.

PASCUAL FERRARA
Sanofi Recherche, 31676 Labège Cedex, France.

J. HAMMER
Hoffmann-La Roche Inc., 340 Kingsland Street, Nutley, New Jersey 07110, USA.

MICHAEL R. JACKSON
R. W. Johnson Pharmaceutical Research Institute, San Diego, CA 92121, USA.

PURNIMA LAUD
Department of Immunology, MD Anderson Cancer Centre, Box 180, 1515 Holcombe Blvd, Houston, TX 77030, USA.

DAVID A. LAWLOR
Department of Immunology, MD Anderson Cancer Centre, Box 180, 1515 Holcombe Blvd, Houston, TX 77030, USA.

ANDREW MELLOR
Institute of Molecular Medicine and Genetics, Medical College of Georgia, 1120, 15th Street, Augusta, GA 30912–3175, USA.

IAN E. G. MORRISON
University of Essex, Wivenhoe Park, Colchester CO4 3SQ, UK.

MARIA PIA PISTILLO
Immunogenetic Laboratory, National Institute for Cancer Research, Largo Rosanna Benzi, 10, 16132 Genova, Italy.

RANDALL K. RIBAUDO
Laboratory of Immune Cell Biology, NCI, National Institutes of Health, Bethesda, Maryland, USA

PHILIPPE SANSÉAU
Glaxo-Wellcome Medicines Research Centre, Gunnels Wood Road, Stevenage SG1 2NY, UK.

F. SINIGAGLIA
Roche Milano Ricerche, Via Olgettina, 58 1-20132, Milano, Italy.

PATRICIA R. SMITH
University of Essex, Wivenhoe Park, Colchester CO4 3SQ, UK.

ANNE-MARIT SPONAAS
Waldstrasse 36, Villiprott 53343, Wachtberg, Germany.

JOHN TROWSDALE
Department of Pathology, University of Cambridge, Tennis Court Road, Cambridge CB2 1QP, UK.

KEITH M. WILSON
University of Essex, Wivenhoe Park, Colchester CO4 3SQ, UK.

Abbreviations

APS	ammonium persulfate
β2-m	β2-microglobulin
BSA	bovine serum albumin
C	cytoplasmic
CBB	Coomassie brilliant blue R-250
CCD	charge-coupled device
DC	dendritic cells
DOC	deoxycholate
DTT	dithiothreitol
EBV	Epstein–Barr virus
ELISA	enzyme-linked immunosorbent assay
EMCV	encephalomyocardis virus
ER	endoplasmic reticulum
ES	embryonic stem
FCS	fetal calf serum
FOF	file-of-file names
FRAP	fluorescence recovery after photobleaching
GCG	Genetics Computer Group
GM	granulocyte/macrophage
GSSG	oxidized glutathione
HAT	hypoxanthine, aminopterin, thymidine
HCG	human chorionic gonadotrophin
IEF	isoelectric focusing
LCL	lymphoblastoid cell line
LDL	low density lipoprotein
LGT	low gelling temperature
mAbs	monoclonal antibodies
MHC	major histocompatibility complex
MLC	mixed lymphocyte culture
MS	mass spectrometry
m.u.	mass units
OTU	operational taxonomic unit
p1	position 1
PAGE	polyacrylamide gel electrophoresis
PBL	peripheral blood lymphocyte
PBS	phosphate-buffered saline
PCR	polymerase chain reaction
PFGE	pulsed-field gel electrophoresis
pI	isoelectric point
PMS	pregnant mare serum

PMSF	phenylmethylsulfonyl fluoride
SCID	severe combined immunodeficiency disease
SDS	sodium dodecyl sulfate
TCR	T cell receptor
TEMED	tetramethylethylenediamine
TFA	trifluoroacetic acid
Tg	transgenic
TGF	Tg founders
TM	transmembrane
YAC	yeast artificial chromosome

1

Generation of HLA-specific human monoclonal antibodies

MARIA PIA PISTILLO and GIOVANNI BATTISTA FERRARA

1. Introduction

HLA class I and class II antigens encoded in the major histocompatibility complex (MHC) are polymorphic membrane glycoproteins whose function is to bind immunogenic peptides inside the cell and transport them to the cell surface, where they can be presented to specific T cells. The interaction between T cell receptor, HLA molecules, and the presented peptides is essential for self and non-self discrimination and for the generation of immune responses.

The earliest approaches used to define the polymorphism of HLA molecules were essentially serological based on the use of conventional (polyclonal) antisera derived from polytransfused patients, multiparous women, skin graft recipients, or from planned immunization of healthy volunteers (1–3). Such sera frequently reacted not only with cells carrying the immunizing HLA molecule, but also with cells carrying antigens controlled by other HLA alleles at the same locus. This finding was mainly due to the antibody complexity for the presence of several antibody populations that recognized different epitopes of the HLA molecule and (or) the same epitope with different affinities. In addition, sequence similarities between HLA alleles contributed to cross-reactivity of sera. When these sera were used to identify the class II antigens, previously identified by cellular methods, they revealed problems of contamination with class I antibodies that required complex analysis and extensive serum absorption.

The monoclonal antibody technology developed by Kohler and Milstein in 1975 helped to solve many of the problems associated with polyclonal antisera in HLA typing as well as in many other areas. Using this technology, monospecific antibodies, homogeneous in immunoglobulin populations and recognizing single antigenic determinants with constant affinity, have been produced. Over the years a large number of monoclonal antibodies (mAbs) against HLA class I and class II molecules have been raised by using the conventional hybridoma procedure of immunizing mice with human material.

However, the majority of the murine anti-HLA mAbs produced are directed to monomorphic determinants of HLA molecules or preferentially to certain HLA specificities (4). In addition, the problem of a frequent murine response to species-specific differences has limited the production of polymorphic anti-HLA mAbs.

In contrast, human anti-HLA mAbs are directed to polymorphic determinants and may even detect HLA allodeterminants that have not yet been identified by either polyclonal alloantisera or murine mAbs. Efforts have therefore been concentrated on methods to generate human mAbs to obtain reagents useful for further dissecting HLA polymorphism and for a therapeutical application in certain HLA-linked diseases.

2. General principles for the generation of human mAbs

The production of human monoclonal antibodies expressing a desired specificity has been tried with methods that use either the Epstein–Barr virus (EBV) to immortalize antigen-specific B cells or the classical hybridoma technique. In this approach human blood lymphocytes are fused with standard mouse myeloma cells, with human lymphoblastoid cell lines (LCLs) or with mouse–human hybrid myelomas, 'heteromyeloma' cells.

EBV is a lymphotropic herpes virus that infects human lymphocytes through the CD21 complement receptor. EBV thus behaves as a polyclonal B cell activator, which allows the expansion *in vitro* of rare antigen-specific B cells from peripheral blood (5). Therefore, B cells from immunized healthy donors can be infected with EBV to generate continuously growing lymphoblastoid cell lines (LCLs) secreting antibodies of predefined specificity. However, resulting cell lines secrete a very low amount of antibody (10–100 ng/ml), have a poor cloning efficiency, and progressively lose antibody production and/or cell proliferation with time. As an alterative, human lymphocytes have been hybridized with mouse myeloma cells resulting in human–mouse hybridomas. These cell lines are generally unstable for the preferential loss of human chromosomes or defects of gene expression. Significant improvement in reducing such stability has been obtained by using 'heteromyelomas' as fusion partners since these cells offer a more favourable environment for retaining the human chromosomes (6). LCLs have also been used as fusion partners but they generate hybrids secreting low levels of antibody (7).

For the production of human monoclonal antibody-secreting cell lines, the combination of the two approaches (i.e. EBV transformation and cell hybridization), is the most widely used method. The strategy developed in our laboratory for the production of human mAbs to polymorphic HLA allodeterminants is based on EBV transformation of HLA immune B lymphocytes, propagation of the antibody-producing cells by 'cluster picking', and fusion of the established EBV-transformed cell lines with an heteromyeloma. Using this procedure we have succeeded in producing several human mAbs recognizing HLA polymorphic class I and class II specificities (8, 9).

2.1 Generation of EBV cell lines secreting anti-HLA mAbs

HLA immunized lymphocytes are purified from peripheral blood (see *Protocol 1*) of either multiparous women or healthy volunteers immunized with allogeneic cells, available in the context of a planned immunization program using whole blood (2) set-up at the Immunohematologic Research Center of Bergamo, Italy. This program is no longer carried out. All the blood donors show a high titre of circulating HLA antibodies and are periodically typed for all officially recognized HLA-A, -B, -C, -DR, and -DQ specificities as defined by cytotoxicity assay, and also for the additional -DP specificities detected by oligotyping. This typing allows us to choose donor–recipient combinations which share some of the HLA specificities and differ in others.

B cell purification is performed by panning with polyvalent anti-human immunoglobulins (see *Protocol 2*) and the purified B cells are exposed to transforming EBV (B95-8 strain) (see *Protocol 3*). After EBV exposure, low density cell cultures (5×10^3 cell/well) are set-up in tissue culture plates using human embryonal fibroblasts as feeder cells (see *Protocol 3*). Clusters of transformed cells appear within two weeks and usually grow to sufficient density for testing in four weeks. The culture supernatant of each well is then tested for the presence of anti-HLA antibodies.

Protocol 1. Isolation of lymphocytes from peripheral blood

Reagents

- Hanks' balanced salt solution (HBSS, Northumbria Biologicals Ltd., M812)
- Ficoll separating solution (Seromed L6115), density 1.077
- Culture medium: RPMI 1640 medium (Seromed, F1215) supplemented with 10% fetal calf serum (FCS, Seromed S0115-1), 2 mM L-glutamine (Seromed K0282), and 100 U/ml penicillin, 100 U/ml streptomycin (Seromed A2213)

Method

1. Dilute 20–50 ml of defibrinated blood from HLA immunized individual 1:4 with Hanks' solution.
2. Carefully layer 10 ml of the above dilution over 4 ml of Ficoll in 15 ml plastic tubes and centrifuge at 800 *g* for 30 min.
3. Collect the resulting interface of mononuclear cells in a 15 ml tube, fill the tube with Hanks' solution, and centrifuge at 800 *g* for 15 min.
4. Discard supernatant, resuspend lymphocytes in 15 ml of complete medium, and centrifuge at 200 *g* for 10 min.

Protocol 1. *Continued*

5. Discard supernatant, resuspend lymphocytes in 2 ml of complete medium and determine total number and cell viability on a cell counting chamber.

Protocol 2. Preparation of B cells by panning and transformation with EBV

Equipment and reagents

- 10 cm squared plastic Petri dishes (Falcon 1001)
- Polyvalent affinity-purified goat anti-human Ig (Cappel 55057)
- EBV-containing supernatant
- Phosphate-buffered saline (PBS, Seromed L1815)
- Flat-bottomed 96-well tissue culture plates (Falcon 3072)

Method

1. Cover the Petri dish with 5 ml of a solution of goat anti-human Ig 10 μg/ml in PBS, and incubate at 37 °C for 2 h.
2. Add 2 ml of FCS to saturate unbound sites and incubate at 37 °C for 30 min.
3. After three washes of the dish with 5 ml of RPMI 1640, cover with 5 ml of the mononuclear cell suspension in culture medium at a maximum rate of 40×10^6 cells/dish.
4. Incubate 3–4 h at 4 °C with occasional agitation to allow complete attachment of B cells by surface Ig.
5. Remove the non-adherent cells (mainly T lymphocytes) by gentle rinsing of the dish three times (be careful to prevent drying out of plate surface).
6. Add 4 ml of EBV supernatant and incubate overnight at 37 °C in a humidified atmosphere of 5% CO_2 in air.
7. The next day detach B cells by repeated washes with complete medium in a 10 ml syringe. Transfer the B cells to a 50 ml tube and count.
8. Centrifuge at 200 *g* for 10 min.
9. Plate the B cells at a final concentration of 5×10^3 cells/well in 96-well flat-bottomed tissue culture plates previously seeded with human fibroblasts as feeder cells. Incubate at 37 °C in a humidified atmosphere of 5% CO_2.
10. Feed EBV cell cultures twice a week with culture RPMI 1640. Feeding is

done by vacuuming off half the medium in a well of the plate and replacing with 100 μl of fresh medium.

11. Screen the cultures for antibody production three to five days after the last feeding.

Protocol 3. Preparation of EBV-containing supernatant and feeder cells

Equipment and reagents

- Marmoset B95-8 cell line (ECACC)
- 75 cm^2 flasks (Greiner 658170)
- 0.45 μm membrane filter (Nalgene 190-2045)
- Trypsin/EDTA solution (Seromed L2143)
- Human embryonic fibroblasts in young passage
- Dulbecco's minimum essential medium (DMEM, Seromed F0435)

A. *EBV supernatant*

1. Culture the marmoset B95-8 cell line in culture medium until saturation for a period of two weeks in 75 cm^2 flasks.
2. Remove the cells by low speed centrifugation at 150 *g* for 10 min.
3. Collect the supernatant and filter through a 0.45 μm membrane filter to ensure removal of viable cells.
4. Store the filtered aliquots of virus at −80 °C and generally use them without further concentration or purification. Do not freeze–thaw aliquots.

B. *Feeder cells*

1. Culture human fibroblasts in 75 cm^2 flasks to confluency in culture DMEM containing 10% FCS and 1% L-glutamine.
2. Remove the supernatant and wash the fibroblast monolayer twice with PBS without calcium and magnesium to remove any FCS trace.
3. Add 4 ml of trypsin/EDTA and incubate at 37 °C for 3 min.
4. Stop the reaction by adding culture DMEM, collect fibroblasts in a 50 ml tube, and centrifuge at 150 *g* for 10 min.
5. Resuspend the fibroblasts and irradiate the suspension under a gamma emitting radiation source (1500 rads) or treat them with mitomycin C.
6. Plate 10^3 fibroblasts/well in 96-well plates and allow to attach overnight at 37 °C.

2.2 Screening of EBV cell culture supernatants

The technique used for testing HLA human mAbs is identical to that used for testing HLA alloantibodies, murine mAbs, and for determining the HLA type. This technique is the complement-dependent microlymphocytotoxicity assay (see *Protocol 4*). Although the technique has many variations the National Institute of Health (NIH) method is generally accepted as the standard procedure. In this laboratory the target cells used in the microlymphocytotoxicity assay to initially screen cell culture supernatants are LCLs obtained from the blood donor volunteer used for HLA-specific immunization. Cytotoxic activity is detected in approximately 2% of wells. Cells from the positive wells are subcultured by the 'cluster picking' technique, and after two to four weeks are again screened for the presence of cytotoxic antibody.

Protocol 4. Microlymphocytotoxicity assay

Equipment and reagents

- 60-well Terasaki trays (Robbins Scientific 1004-00-0)
- Hamilton microdispenser (Robbins Scientific 1023-01-0) and multiple syringe (Robbins Scientific 1021-01-0)
- Coverslides (Terasaki Insta-Seal coverslides, One Lambda TIS-250)
- Rabbit complement (One Lambda CABC-5)

Method

1. Dispense 1 μl of test mAb into each well of the Terasaki tray (previously seeded with one drop of mineral oil) using a Hamilton syringe. Also dispense positive and negative controls.
2. Add 1 μl of target cells (EBV cells, B cells, or PBLs) (3×10^6 cell/ml in RPMI 5% AB serum) into each well and check that the cell suspension and supernatants have mixed.
3. Incubate at 20°C for 60 min in the case of EBV or B cell targets, and for 30 min in the case of PBLs or T cell targets.
4. Add 5 μl of rabbit complement (previously tested for toxicity and activity on the target cells) to each well with a multiple syringe.
5. Incubate at 22°C for 2 h in the case of EBV or B cell targets, and for 60 min in the case of PBLs or T cell targets.
6. Add 3 μl of 5% aqueous eosin to each well and after 3 min 9 μl of fixative (formaldehyde adjusted to pH 7.4 with $NaHCO_3$).
7. Carefully lower a coverslide on wells and incubate overnight at 4°C.
8. Determine the per cent of cell lysis under an inverted phase-contrast microscope. Viable cells appear bright and refractile whereas dead

cells appear larger than live cells, dark, and non-refractile. Per cent of cell death is recorded using the following International Histocompatibility Workshop scoring system: 1 = negative (0–10%), 2 = doubtful negative (11–20%), 4 = weak positive (21–40%), 6 = positive (41–80%), 8 = strong positive (81–100%).

2.3 Cluster picking

Because of the low cloning efficiency of EBV cells by traditional cloning procedure (limiting dilution or growth in semi-solid medium) we developed the 'cluster picking' technique (10) (see *Protocol 5*) as a measure to subculture the antibody-producing cells and perpetuate antibody production. This technique is based on picking up individual clusters of cells from the cytotoxicity positive wells and transferring them to new wells with the aid of a stereoscopic microscope. After two to four weeks, the new cultures are again screened for the presence of cytotoxic antibody. The antibody-secreting cells are in this way selected and the non-secreting ones eliminated. Thus, the sublines established are enriched in cells producing antibodies of HLA specificity, and undoubtedly this increases the chance of successful formation and isolation of hybrids producing the desired antibody upon fusion with a suitable partner.

The subculture by 'cluster picking' is repeated at least three times and after that uniformly antibody positive cells are observed.

Protocol 5. Cluster picking technique

Equipment

- Stereoscopic microscope (Leitz Wild M8)
- Curved Pasteur pipetter tip

Method

1. Transfer clusters of cells from the antibody positive cultures to new wells of a 96-well plate containing human fibroblasts. This procedure is performed with the aid of a curved Pasteur pipetter tip used under a stereoscopic microscope placed into a laminar flow-hood.
2. Incubate the new plates at 37 °C for two to three weeks feeding twice a week with culture medium.
3. When the cultures are grown to the desired density, test them for antibody production.
4. Repeat 'cluster picking' from the positive wells.
5. During 'cluster picking' passages collect supernatants (antibody) and freeze cells in liquid nitrogen.

2.4 Monoclonality determination of anti-HLA human mAbs

The monoclonality of the human antibody secreted by the established EBV cell lines after 'cluster picking' propagation is evaluated by testing its immunoglobulin isotype, HLA monospecificity, and cell monoclonality. Since EBV cells have a poor cloning efficiency they must be hybridized to generate hybrid cells that can be easily cloned at less than one cell per well and can therefore be regarded as monoclonal cells.

The immunoglobulin isotype of human mAbs can be determined by cytoplasmic immunofluorescence staining of the antibody-secreting cells using human chain-specific antibodies (see *Protocol 6*), or by ELISA technique of culture supernatants using enzyme-linked anti-human Ig antibodies (see *Protocol 7*).

Protocol 6. Cytoplasmic immunofluorescence

Equipment and reagents

- 8-well multitest slides (Flow 60-408-05)
- Poly-L-lysine (Sigma P5899)
- 37% formaldehyde solution (Merk 4002)
- Triton X-100 (Bio-Rad 161-0407)
- FITC-conjugated anti-human IgM, IgG, *k*, and λ chains (Cappel 55185, 55184, 55188, and 55189)

Method

1. Dispense 50 μl of poly-L-lysine (0.1 mg/ml in PBS) into each well of 8-well multitest slides and incubate at room temperature for 30 min.
2. Wash three times with distilled water by vacuuming off the water and replacing it. Dry.
3. Add 5×10^4 cells in 30 μl of PBS into each well and incubate at room temperature for 20 min.
4. Remove PBS and add 20 μl of 3.7% formaldehyde in PBS. Incubate at room temperature for 5 min.
5. Wash three times with PBS and add 20 μl of 0.05% Triton X-100 in PBS. Incubate at room temperature for 5 min.
6. Wash three times with PBS containing 1% FCS.
7. Add 20 μl of FITC-conjugated anti-human IgM and IgG antibodies (1:80 in PBS, 1% FCS) and anti-human *k* and λ chains (1:100 in PBS, 1% FCS). Incubate at room temperature for 20 min.
8. Wash six times with PBS.
9. Mount the slides with coverslides and analyse the percentage of stained cells under a fluorescence microscope (Leitz).

Protocol 7. ELISA

Equipment and reagents

- 96-well plates (Nunc 54394541)
- Rabbit anti-human Ig *k* and λ light chains (Dako M730, M614)
- Alkaline phosphatase-conjugated goat anti-human Ig (Cappel 59284)
- *P*-Nitrophenylphosphate disodium (substrate tablets, Sigma N9389)
- Bovine serum albumin (BSA, Sigma A2153)
- Tween 20 (Sigma P1379)

Method

1. Coat plates with 100 μl of rabbit anti-human light chains (1:10 in PBS) and incubate at 37 °C for 2 h.
2. Flick and wash three times with PBS containing 0.05% Tween 20.
3. Block the plate by adding 200 μl of 3% BSA in PBS and incubate overnight at 4 °C.
4. Flick and wash three times with PBS, 0.05% Tween 20.
5. Add 100 μl of test supernatants plus negative (PBS) and positive (IgG and IgM at known concentrations) controls.
6. Flick and dry the plates by hammering them face down on wads of paper.
7. Add 100 μl of alkaline phosphatase-conjugated goat anti-human Ig 1:300 in PBS and incubate at 37 °C for 1 h.
8. Wash four times with PBS, 0.05% Tween 20.
9. Add 100 μl of substrate prepared by dissolving one tablet (5 mg) in 5 ml of diethanolamine buffer pH 9.8 (48 ml diethanolamine and 24.2 mg $MgCl_2$ in 500 ml final volume with bidistilled water).
10. Stop reaction with 3 M NaOH.
11. Read the absorbance at 492 nm in a spectrophotometer such as the Titertek Multiskan (Flow Laboratories).

2.5 Fusion and cloning of hybridomas

Once EBV-transformed cells are stabilized they can be fused to a suitable HAT-sensitive and ouabain-resistant fusion partner (see *Protocol 8*) in order to increase:

- antibody production
- cloning efficiency
- stability in growth and Ig secretion

In fact, human/mouse hybridomas secrete higher levels of Ig compared to the parental EBV cell line, can be easily cloned at less than one cell per well, do not stop growing, and retain antibody secretion for longer periods.

Table 1. Currently used fusion partners in human monoclonal antibody production

Fusion partner	Cell type	Secreted Ig
SKO-007	Myeloma	IgE (λ)
RPMI 8226	Myeloma	λ
HFB 1	Myeloma	None
KMMI	Myeloma	G
RH-L4	Lymphoma	
NAT-30	Lymphoma	
GM 1500-6TG-AL	LCL	IgG2 (*k*)
GM 1500-6TG-OA	LCL	
KR4	LCL	IgG2 (*k*)
GM 4672	LCL	IgG2 (*k*)
ARH-77	LCL	IgG (*k*)
LICR-LON-HMY2	LCL	IgG1 (*k*)
WI-L2	LCL	
MC/CAR	LCL	None
MC/MNS2	LCL	IgG1 (*k*)
LTR 228	LCL	IgM (*k*)
LSM 2-7	LCL	
HS Sultan	LCL	None
GK-5	LCL	*k*
HO-323	LCL	
KR-12	Heteromyeloma	IgG2 (*k* + λ)
SHM-D33	Heteromyeloma	None
SBC-H20	Heteromyeloma	None
3 HL	Heteromyeloma	IgM (λ)

Unfortunately, the ideal fusion partner, a human myeloma cell line that should give karyotypically stable intraspecies hybrids, is not available at this time and this has encouraged the use of other fusion partners. The partner lines used are of different phenotypes (see *Table 1*) including myelomas that are not frequently used, mainly because they grow very poorly in culture. The majority of human fusion partners are LCLs, derived by EBV transformation of lymphocytes, that grow better than myelomas, but generate hybridomas secreting low levels of antibody and usually mixed immunoglobulin molecules derived from both fusion partners. As an alternative fusion partner, heteromyelomas have been constructed by the fusion of murine myeloma cells with human cells. Most heteromyelomas produced so far have proven to generate high yields of viable hybridomas, sustain secretion of relatively high levels of human immunoglobulins, and allow higher stability of immunoglobulin secretion. The fusion partner used in this laboratory is the non-Ig-secreting human/mouse heteromyeloma SHM-D33 (6).

In order to allow adequate selection of hybrid cells all the fusion partners must be HAT-sensitive and ouabain-resistant because the EBV cells are already immortalized and they can be eliminated by the use of ouabain. The fusion partners are HAT-sensitive and hence lack the enzyme hypoxanthine guanine phosphoribosyl transferase (HGPRT) that is required in the salvage pathway of nucleotide biosynthesis. Therefore, they are not capable of growing when the main pathway of DNA synthesis is blocked with HAT medium. Hybrid cells can grow in culture medium supplemented with HAT because the EBV cells provide the necessary enzyme HGPRT. Since the fusion partners, and as a consequence the hybrid cells, are ouabain-resistant, addition of this drug kills only unfused EBV cells.

Protocol 8. Fusion with SHM-D33 and cloning by limiting dilution

Reagents

- Heteromyeloma SHM-D33 (ATCC)
- 50% PEG 1500 (Boehringer Mannheim 783641)
- Peritoneal macrophages from Balb/c mice
- Hypoxanthine, aminopterin, thymidine (HAT, Flow 16-808-49), hypoxanthine, thymidine (HT, Flow 16-809-49), and ouabain (Sigma 0-3125)

Method

1. Harvest EBV cells and SHM-D33 myeloma separately from the logarithmic phase of growth at $> 90\%$ cell viability and centrifuge twice in 50 ml of serum-free RPMI 1640 at 150 *g* for 10 min.
2. Mix the two cell lines at a ratio of 2:1 (excess of EBV cells) in 20 ml of serum-free RPMI 1640 in 50 ml tube and centrifuge twice at 150 *g* for 10 min.
3. Gently disrupt pellet after removal of the supernatant and add 0.4 ml of 50% PEG 1500, pre-warmed to 37 °C, drop by drop over 45 sec.
4. Add 10 ml of complete medium slowly over 90 sec and then other 20 ml over 3–5 min.
5. Incubate the fused cells in water-bath at 37 °C for 5 min.
6. Centrifuge at 150 *g* for 4 min. Resuspend in RPMI 1640, 20% FCS and plate at 10×10^3 cells/well in 96-well tissue culture plates with peritoneal macrophages (10×10^3 cells/well). Also set-up control cultures of each parental cell line to check the surviving cells in selective medium.
7. The next day remove medium and add culture RPMI 1640 plus 2% HAT and 10^{-7} M ouabain. Feed twice a week with this medium for a period of 15 days, and then feed with culture RPMI 1640 plus 2% HT for other 15 days. Then feed only with culture RPMI 1640.
8. Screen each well for antibody production.
9. Clone positive wells by limiting dilution plating on peritoneal macrophages in serial, tenfold dilutions from 10 up to 0.5 cells/well in 96-well flat-bottomed plates.

2.6 Growth of human hybridomas in immunodeficient mice

It is well established that high yields of monoclonal antibodies produced by mouse hybridomas can be obtained by growing these cells as ascites tumours in pristane-treated syngeneic mice. However, hybridomas of human origin are not easily transplanted into mice because of non-self immune recognition. This problem can be partially solved by suppression of the immune system by drugs or irradiation but these methods may compromise the growth of the hybridoma. The use of athymic, nude mice, for ascites production has been shown useful for human hybridomas, but these animals still require irradiation and *in vivo* adaptation of the hybridomas. In addition, the antibody produced in the ascites can be contaminated by endogenous immunoglobulin molecules. There have been a number of successful attempts to grow human hybrids in mice with severe combined immunodeficiency disease (SCID). These mice bear a spontaneous autosomal recessive mutation responsible for a deficiency in both B and T lymphocytes whereas natural killer activity is fully maintained. Results from this laboratory (11) show that the level of human mAb in ascites produced by a human/mouse hybridoma can be increased approximately 100-fold also resulting in increased specific cytotoxicity activity.

3. Analysis of HLA specificity recognized by human mAbs

The HLA specificity of the human mAb produced is generally studied by testing the antibody reactivity on a panel of well characterized cell lines such as HLA homozygous LCLs (12). These lines have been serologically typed in successive workshops and more recently sequenced for the HLA class II alleles, DRB, DQA1, DQB1, DPA1, and DPB1. The panel should contain cells covering all the different HLA antigens and if possible several cells expressing the same antigen. Segregation analysis within informative families should also be carried out. Once the specificity has been assigned this must be confirmed by testing the mAb reactivity against peripheral B lymphocytes (for class II antigens) and PBLs (for class I antigens) since these are the cells generally used in a routine tissue typing laboratory. These LCLs may be used in cytotoxicity assay or flow cytometry (see *Protocol 9*). To more precisely identify the HLA specificity recognized by the human mAb, murine target cells transfected with different human HLA class I and II genes have been developed. Since these transfectants express on their cell surface a single HLA class I or II allele, they are useful tools not only to determine locus and α or β chain specificity but also to map single amino acid residues or clusters of residues involved in the antibody binding site. These transfectants may be used in flow cytometry or microcell enzyme-linked immunosorbent assay (mirocell ELISA) (see *Protocol 10*).

Protocol 9. Flow cytometry

Equipment and reagents

- 5 ml polypropylene tubes (Falcon 2063)
- Fluorescein (FITC)-conjugated goat anti-human total Ig (Cappel 55186)
- Flow cytometry, e.g. FACStar (Beckton Dickinson)

Method

1. Incubate 100 μl of human mAb with an equal volume of cell suspension (1×10^5 cells) in 5 ml tubes at 4 °C for 30 min. Also set-up positive and negative controls.
2. Wash twice in PBS by centrifugation at 200 *g* for 5 min, incubate with FITC-conjugated anti-human Ig (1:20 in PBS) for 30 min at 4 °C.
3. Wash three times in PBS.
4. Resuspend the cells in 200 μl medium and analyse using FACS equipment.

Protocol 10. Microcell ELISA

Equipment and reagents

- Glutaraldehyde (Sigma G-6257), casein (Sigma C-0376)
- Peroxidase-linked anti-human Ig (Cappel 55248)
- Micro EIA plate reader (Genetic Systems)
- 2,2′azino-*bis*(3-ethylbenzthiazoline-6-sulfonic acid) (ABTS, Sigma NA 1888)
- pH 5 buffer: 0.1 M citrate buffer, 0.1 M disodium hydrogenophosphate

Method

1. Dispense 1×10^4 viable transfectant cells in a 10 μl suspension volume in each well of a Terasaki plate and incubate for 24 h at 37 °C.
2. Flick out before use and wash the plates with cells attached twice with PBS. Blot on tissue paper.
3. Add 10 μl of 0.025% glutaraldehyde in PBS to each well. Stand at room temperature for 15 min.
4. After two washes with PBS, 0.2% casein, add 5 μl of test mAb to each well and incubate at room temperature for 60 min. Wash again with PBS, 0.2% casein.
5. Add 5 μl of peroxidase-linked anti-human Ig (1:300 in PBS) and incubate at room temperature for 60 min.
6. Wash five times with PBS, 0.2% casein and add 5 μl of substrate (prepared by dissolving 5 mg of ABTS in 10 ml of pH 5 buffer plus 5 μl 30% hydrogen peroxide).
7. Read the optical density at 660 nm on a Micro EIA plate reader after 10–30 min at room temperature in the dark.

Table 2. DR beta chain first domain amino acid sequences

```
           1        10        20        30        40        50        60        70        80
           .        .         .         .         .         .         .         .         .
DRB1*0101  GDTRPRFLWQLKFECHFFNGTERVRLLERCIYNQEESVRFDSDVGEYRAVTELGRPDAEYWNSQKDLLEQRRAAVDTYCRHNYGVGESF
DRB1*0102  ------------------------------------------------------------------------------------AV---
DRB1*0103  ------------------------------------------------------------------I--DE------------------
DRB1*1501  ----------P-R------------F-D-YF---------------F-------------------I---A--------------V---
DRB1*1502  ----------P-R------------F-D-YF---------------F-------------------I---A-----------------
DRB1*1503  *****-----P-R------------F-D-HF---------------F-------------------I---A--------------V---
DRB1*1601  ----------P-R------------F-D-YF-----------------------------------F--D-------------------
DRB1*1602  ----------P-R------------F-D-YF--------------------------------------D-------------------
DRB1*0301  --------EYSTS------------Y-D-YFH----N---------F-----------------------K-GR--N--------V---
DRB1*0302  --------EYSTS------------F---YFH----N---------------------------------K-GR--N------------
DRB1*0401  --------E-V-H------------F-D-YF-H---Y---------------------------------K------------------
DRB1*0402  --------E-V-H------------F-D-YF-H---Y-----------------------------I--DE--------------V---
DRB1*0403  --------E-V-H------------F-D-YF-H---Y------------------------------------E-----------V---
DRB1*0404  --------E-V-H------------F-D-YF-H---Y------------------------------------------------V---
DRB1*0405  --------E-V-H------------F-D-YF-H---Y-------------------S--------------------------------
DRB1*0406  --------E-V-H------------F-D-YF-H----------------------------------------E-----------V---
DRB1*0407  --------E-V-H------------F-D-YF-H---Y------------------------------------E---------------
DRB1*0408  *************------------F-D-YF-H---Y----------------------------------------------------
DRB1*1101  --------EYSTS------------F-D-YF-----Y---------F----------E--------F--D----------
DRB1*1102  --------EYSTS------------F-D-YF-----Y---------F----------E--------I--DE--------------V---
DRB1*1103  --------EYSTS------------F-D-YF-----Y---------F----------E--------F--DE------------------
DRB1*1104  --------EYSTS------------F-D-YF-----Y---------F----------E--------F--D---------------V---
```

```
DRB1*1201  --------EYSTG--Y-------------HFH----LL--------F---------V--S------I--D--------------AV---
DRB1*1301  --------EYSTS------------F-D-YFH----N---------P-------------------I--DE--------------V---
DRB1*1302  --------EYSTS------------F-D-YFH----N---------P-------------------I--DE------------------
DRB1*1303  --------EYSTS------------F-D-YF-----Y-------------------S---------I--DK------------------
DRB1*1401  --------EYSTS------------F-D-YFH----F-------------------A--H---------R---E-----------V---
DRB1*1402  --------EYSTS------------F---YFH----N----------------------------------------------------
DRB1*1403  --------EYSTS------------F---YFH----N--------------------------------D---L---------------
DRB1*0701  ---Q------G-YK----------QF---LF-----F-------------------V--S------I--D--GQ---V-----------
DRB1*0702  ---Q------G-YK----------QF---LF-----F-------------------V--S------I--D--GQ---V-----------
DRB1*0801  --------EYSTG--Y---------F-D-YF-----Y-------------------S---------F--D---L-------
DRB1*0802  --------EYSTG--Y---------F-D-YF-----Y-----------------------------F--D---L---------------
DRB1*0803  *****---EYSTG--Y---------F-D-YF-----Y-------------------S---------I--D---L---------------
DRB1*0804  --------EYSTG--Y---------F-D-YF-----Y-----------------------------F--D---L-----------V---
DRB1*0901  ---Q----K-D--------------Y-H-G------N-------------------V--S------F--R---E---V-----------
DRB1*1001  --------EEV------------------RVH----YA-Y---------------------------------R---------------
DRB3*0101  --------ELR-S------------Y-D-YFH----FL------------------V--S----------K-GR--N------------
DRB3*0201  --------EL--S------------F---HFH----YA-------------R------------------K-GQ--N--------V---
DRB3*0202  --------EL--S------------F---HFH----YA-------------R------------------K-GQ--N------------
DRB3*0301  --------EL--S------------F---YFH----F-------------------V--S----------K-GQ--N--------V---
DRB4*0101  ---Q----E-A-C----L------WN-I-Y------YA-YN--L---Q---------------------R---E------Y----Y---
DRB5*0101  --------Q-D-Y------------F-H-D------DL----------------------------F--D-------------------
DRB5*0102  --------Q-D-Y------------F-H-G------N-----------------------------F--D-------------------
DRB5*0201  -----C--Q-D-Y------------F-H-G------N-----------------------------I---A-------------AV---
DRB5*0202  -----C--Q-D-Y------------F-H-G------N-----------------------------I---A-------------AV---
```

3.1 Identification of the epitope recognized by anti-HLA human mAbs

Nucleotide (amino acid) sequences (13) of the HLA class I and II alleles allow one to correlate the reactivity pattern of the human mAbs with the presence of specific amino acid residues, and therefore to define the epitope which is recognized by the human mAbs. This is done by comparing all the polymorphic sequences that are present in the alleles reacting with the mAb but are absent in all other alleles. An example of epitope mapping of HLA-DR specific human mAbs is shown below.

The human mAb HMP12 reacts with all cells in a panel expressing the alleles DRB1*1101, 1102, 1103, 1104, and the human mAb HMP14 reacts with all cells expressing the alleles DRB1*0801, 0802, 0803, 0804, and DRB1*1201. No reactivity is found against cells expressing different DRB1* alleles. A comparison of the HLA-DRB first domain amino acid sequences (see *Table 2*) shows that the DRB1*1101, 1102, 1103 and 1104 are the only alleles that show the amino acid residue E in position 58, and that the DRB1*0801, 0802, 0803, 0804, and DRB1*1201 are the only alleles that share regions of sequence identity from position 13 to position 16 (GECY). These findings indicate that these residues are involved in the formation of the antibody binding epitopes. The availability of the methods of site-directed mutagenesis and DNA-mediated gene transfer now make it possible to precisely identify these specific amino acid residues.

4. Application of anti-HLA human mAbs

Since human mAbs are generally directed to polymorphic HLA molecules, they can help in understanding the molecular mechanism of processes in which HLA polymorphism plays a critical role such as graft rejection, antigen presentation, and susceptibility to develop certain HLA-linked autoimmune diseases including insulin-dependent diabetes, coeliac disease, and rheumatoid arthritis. In such autoimmune disorders human anti-HLA mAbs can provide useful therapeutic reagents since, at variance with murine mAbs, they do not induce sensitization or allergic reactions upon repeated *in vivo* injections. A striking characteristic of human autoimmune disease is the increased frequency of certain HLA class II alleles in affected individuals. It is possible that such alleles have the capacity to bind autoantigens and present them to T cells thus inducing and maintaining the autoimmune disease. One possible approach to interrupting this phenomenon is the administration of HLA allele-specific human mAbs that should act by blocking the antigen-presenting activity of disease-associated HLA class II molecules avoiding activation of the class II restricted autoreactive T cells.

Acknowledgements

We would like to thank Stefano Mantero and Giovanni De Pascalis for assistance in preparing the chapter.

References

1. Payne, R. (1957). *Arch. Intern. Med.*, **99**, 587.
2. Ferrara, G. B., Tosi, R., Longo, A., Castellani, A., and Carminati, G. (1978). *Transplantation*, **26**, 150.
3. Ceppellini, R., Curtoni, E. S., Mattiuz, P. L., Leigheb, G., Visetti, M., and Colombi, A. (1966). Ann. *N.Y. Acad. Sci.*, **129**, 421.
4. Sachs, J., Fernandez, N., Kurpisz, M., Okoye, R., Ogilvie, R., Awad, J., *et al.* (1986). *Tissue Antigens*, **28**, 199.
5. Steinitz, M., Klein, G., Koskimies, S., and Makela, O. (1977). *Nature*, **269**, 420.
6. Teng, N. N. H., Lam, K. S., Riera, F. C., and Kaplan, H. S. (1983). *Proc. Natl. Acad. Sci. USA*, **80**, 7308.
7. Kozbor, D., Lagarde, A. E., and Roder, J. C. (1982). *Proc. Natl. Acad. Sci. USA*, **79**, 6651.
8. Pistillo, M. P., Mazzoleni, O., Lu, K., Falco, M., Tazzari, P. L., and Ferrara, G. B. (1991). *Hum. Immunol.*, **31**, 86.
9. Mazzoleni, O., Pistillo, M. P., Falco, M., Tazzari, P. L., and Ferrara, G. B. (1991). *Tissue Antigens*, **38**, 224.
10. Pistillo, M. P., Hammerling, U., Dupont, B., and Ferrara, G. B. (1986). *Hum. Immunol.*, **15**, 109.
11. Pistillo, M. P., Sguerso, V., and Ferrara, G. B. (1992). *Hum. Immunol.*, **35**, 256.
12. Yang, S. Y., Milford, E., Hammerling, U., and Dupont, B. (1989). In *Immunobiology of HLA* (ed. B. Dupont), Vol. 1, pp. 11–19. Springer-Verlag, New York.
13. Marsh, S. and Bodmer, J. G. (1992). *Hum. Immunol.*, **35**, 1.

2

Sequencing of proteins isolated by one- or two-dimensional gel electrophoresis

PASCUAL FERRARA

1. Introduction

One-dimensional (1D), and especially two-dimensional (2D) polyacrylamide gel electrophoresis (PAGE), are generally considered as the highest resolving techniques for the analysis of complex protein mixtures. Both techniques have been extensively used for the study of cell protein changes in biological processes such as proliferation and differentiation (1). In 2D PAGE the power of electrophoresis analysis is enhanced when individual spots on the gel are identified. However, the identification of the spots is not a trivial task; several direct and indirect methods have been developed (2, 3). A straightforward approach is the production of sequence information. Such information can then be used, for example, for the design of oligonucleotide probes for the cloning of new genes, or the development of 2D gel databases that may help the interpretation of the DNA sequence data of the human genome (4, 5). N terminal sequence information can be obtained from proteins electroblotted from the gels onto inert supports (6, 7). However, since many proteins are not susceptible to Edman degradation due to either natural or artefactual N terminal blocking, methods have been developed for the digestion of electrotransferred proteins directly on the inert support, and for the elution of the resulting peptides for internal sequence analysis (8, 9). Elution of the electrotransferred proteins prior to digestion has also been described (10, 11), but the variable transfer efficiency of different proteins (12), added to low recoveries of the eluted material, results in significant losses. For these reasons, methods that circumvent electrotransfer, such as passive (13) or electrophoretic (14) elution, are being explored. Proteins recovered from stained gels require further purification before digestion to remove the dye, sodium dodecyl sulfate (SDS), and gel-related contaminants that interfere with the digestion and with the subsequent peptide separation. Selective precipitation (15) or, more recently, inverse gradient reverse-phase HPLC (16) have been used to purify the eluted proteins, but when working with picomolar levels of proteins these multistep procedures result in poor yields (17).

Direct digestion of Coomassie brilliant blue R-250 (CBB) stained proteins in the gel matrix (in-gel digestion) is a very attractive option. However, there are only a limited number of reports that describe in-gel digestions. This may be because the existing methods involve very long and laborious sample treatments before peptide sequencing, during which loss of materials cannot be excluded.

Recently, we have developed a simple method for enzymatic digestion of proteins in the polyacrylamide matrix (18, 19). We describe here in detail the protocols followed in our laboratory to obtain internal amino acid sequence data from proteins isolated by 1D PAGE or preparative 2D gels. These protocols can be applied to the analysis of virtually any protein including MHC products. We also discuss the application of mass spectrometry analysis of the peptides resulting from the in-gel digestion, for the rapid identification of proteins described in the databases.

2. Electrophoresis techniques for the isolation of proteins

The general strategy followed in our laboratory to identify proteins isolated by SDS–PAGE or preparative 2D gels is outlined in *Figure 1*. The choice of SDS–PAGE or 2D gels to prepare the samples for protein characterization depends on the nature of the problem and the complexity of the starting material. In general, 1D gel separations are relatively straightforward, require minimal equipment, and are easily reproducible from laboratory to laboratory. However, when 1D gels are not sufficient to isolate the proteins of interest, either because several proteins co-migrate or because of the complexity of the starting material, high-resolution 2D gels should be used, but the time, effort, and expertise required for this technique are much greater.

2.1 SDS–PAGE

1D SDS–PAGE were prepared using the discontinuous buffer system described by Laemmli (20). The experimental details are given in *Protocol 1*, and for a complete introduction and discussion, as well as a troubleshooting guide, see for example ref. 14. The materials used are: glycine and Tris base from Reidel-deHaen; urea from Schwartz-Mann; SDS, acrylamide, and other reagents for SDS–PAGE and 2D gels from Bio-Rad; CBB, porcine trypsin, formaldehyde, and ammonium carbonate from Sigma. All organic solvents were from SDS (Peypen-France). IL-6 was produced in *E. coli* and purified to homogeneity by HPLC. Molecular weight standard proteins were from Pharmacia. Radiolabelled molecular weight standard proteins were from Amersham. The gels were run either on a Protean II (Bio-Rad), or on an Investigator 2D (Millipore) electrophoresis apparatus.

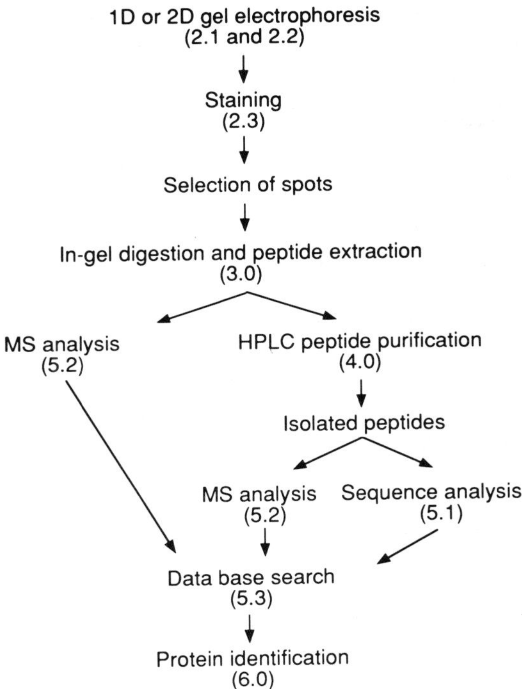

Figure 1. Outline of the strategy for the identification of proteins isolated by gel electrophoresis. The numbers correspond to the sections in the chapter where the steps are described and discussed.

Protocol 1. SDS–PAGE separation of proteins (discontinuous buffer system)

Wear gloves and mask when handling acrylamide powder, and gloves when handling unpolymerized gel solutions, since both are neurotoxic.

Equipment and reagents

- Glass plates
- Solution A: 30% acrylamide, 0.8% *bis*-acrylamide
- Solution B: 0.5 M Tris–HCl pH 6.8 (stacking gel buffer)
- Solution C: 3 M Tris–HCl pH 8.8 (resolving gel buffer)
- Solution D: 20% SDS
- Solution E: 15% ammonium persulfate (APS) (prepare just before use)
- Solution F: 25 mM Tris, 0.192 M glycine pH 8.6, and 0.1% SDS (electrode buffer)
- Solution G: 0.125 M Tris–HCl pH 6.8, 4% SDS, 20% glycerol, 0.001% bromophenol blue (solubilization buffer)
- TEMED: *N,N,N′,N′*-tetramethylethylenediamide
- β-mercaptoethanol

Protocol 1. *Continued*

Method

1. Clean the glass plates with detergent, rinse well with water and with ethanol. Assemble a glass plate sandwich (this protocol is for 18 × 16 cm plates compatible with Bio-Rad Protean II electrophoresis units), sealed on both sides and at the bottom, using spacers of 0.75 mm thickness.
2. For a 12.5% acrylamide running gel combine solutions A: 12.5 ml, C: 3.75 ml, and H_2O: 13.4 ml. Filter (0.2 μm) and degas, then add 150 μl of solutions D (SDS) and E (APS), and 15 μl TEMED. APS and TEMED are added just before pouring out the mixture between the glass plate sandwich. Do not completely fill the glass plate sandwich with the resolving gel solution. 5–8 cm have to be left for the stacking gel.
3. Gently cover with water-saturated isobutanol to protect the gel from adverse effects of oxygen on the polymerization, and to obtain the top of the gel straight.
4. Store overnight with the top of the sandwich wrapped in plastic film to prevent evaporation and, before adding the stacking gel, wash the gel surface with deionized water.
5. Prepare the stacking gel solution with solution A: 1.9 ml, solution B: 3.8 ml, and deionized water: 9.3 ml. Filter, degas, and add 75 μl of solution D. Add 75 μl of solution E and 10 μl of TEMED, mix, and layer the stacking gel solution on top of the resolving gel.
6. Insert a comb of the desired number of wells. Avoid trapping bubbles on the bottom of the comb dividers. Allow to polymerize for at least 1 h at room temperature.
7. Prepare tank buffer. Assemble the gel in the chamber, remove bubbles trapped at the bottom of the gel, and attach and fill with buffer (solution F) the upper buffer reservoir.
8. Prepare the samples: add solution G, 100 μl for 50–100 μg protein, and β-mercaptoethanol, 5 μl.
9. Heat samples for 5 min at 95 °C before loading them into the wells.
10. Gently underlayer the samples in wells. In a terminal well load the molecular weight markers (treated like samples).
11. Electrophoresis can be run at constant intensity (25 mA for 0.75 mm thick slab gel), and at constant temperature (9–10 °C).
12. Stop electrophoresis when the blue migration front reaches the bottom of the gel.
13. Stain the gels as described in *Protocol 4*.

2.2 Two-dimensional electrophoresis

High-resolution 2D gels, as developed by O'Farrel (21), involves the separation of proteins in the first dimension according to their isoelectric point (pI) and in the second dimension according to molecular weight. In general, 1D PAGE and 1D isoelectric focusing are each capable of resolving approximately 100 components of a very complex protein mixture. Combining both techniques in a 2D analysis one can resolve 1000–2000 components in a single gel. Since several excellent and improved protocols have been described to run high-resolution 2D gels (22–24), we will describe only the preparative technique that we use to isolate proteins for characterization. *Protocol 2* describes the preparation and running of the first dimension, and *Protocol 3* describes the preparation and running of the second dimension. These preparative 2D gels can be used to separate, with adequate resolution, up to 2 mg of protein mixture in a single run (*Figure 2*). Basically, we run a preparative isoelectric focusing gel in a 40 × 0.3 cm (internal diameter) glass tube and then we cut it into two 20 cm pieces that are run separately for the second dimension, on 24 cm × 22 cm × 1.5 mm thick polyacrylamide slab gels (this plate size is compatible with the Investigator 2D (Millipore) electrophoresis apparatus). For the materials see Section 2.1. Ampholytes were from Serva (pH 2–11) and Pharmacia (pH 3–10, 4–6.5, and 5–8).

To get an estimate of the complexity of the sample it is useful to run the same sample in a one-dimensional lane next to the isoelectric focusing gel

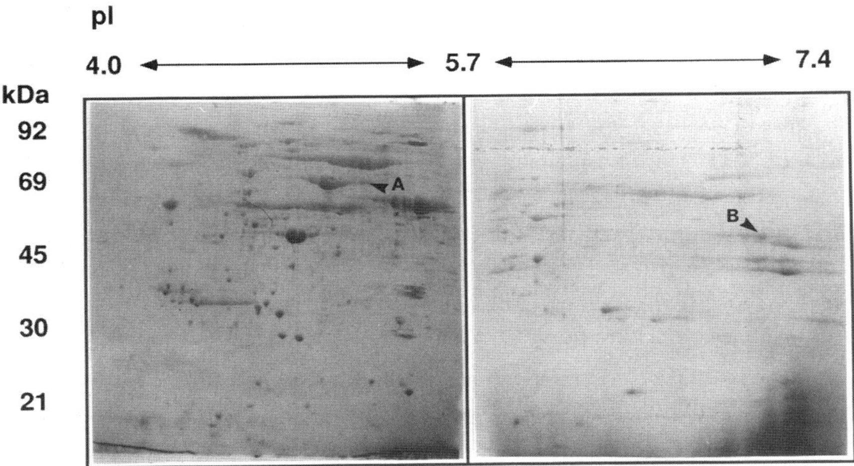

Figure 2. Two-dimensional gel electrophoresis of mitochondrial rat proteins. The first (pI 4.0–5.7)) and second (pI 5.7–7.4) halves of the first dimension were run on two slab gels as described in Section 2.2. Spots A and B were cut out and in-gel digested, the results are shown in *Figures 6* and *7*.

when running the second dimension as described in *Protocol 3*. This may allow the detection of interesting proteins that are out of the pH range used for the first dimension, and will help to find out the most appropriate pH range to run the first dimension.

Protocol 2. Preparative 2D gels: first dimension, isoelectric focusing (IEF)

Equipment and reagents

- 40 × 0.3 cm (internal diameter) glass tubes
- The isoelectric focusing apparatus is home-made in Plexiglas with the following dimensions: 45 × 25 × 11 cm
- Solution A: 30% acrylamide, 0.8% *bis*-acrylamide
- Solution B (ampholytes mixture): 510 μl of pH 3–10, 1020 μl of pH 4–6.5, 1530 μl of pH 5–8, and 510 μl of pH 2–11 — this mixture gives a pH gradient of 4.4 to 7.1 (pH ranges can be modified according to the samples and needs of investigator)
- Solution C: 10% ammonium persulfate (freshly prepared)
- Solution D, gel overlay solution (GOS): 1% ampholytes pH 5–8, 5% CHAPS, 0.1 M DTT, 4 M urea — filter (0.2 μm), aliquot (1 ml), and store at −20 °C
- Solution E (sample buffer): 2% ampholytes pH 5–8, 2% CHAPS, 0.1 M DTT, 9.5 M urea, 0.001% bromophenol blue, and deionized water up to 50 ml — filter, aliquot (1 ml), and store at −20 °C
- Nonidet NP-40
- TEMED
- Urea

A. *Procedure for casting IEF gels*

1. Prepare the IEF gel solution at 4% acrylamide with solution A: 11.7 ml, urea: 51.3 g (final concentration 9.5 M), Nonidet NP-40: 1.9 ml, and deionized water: up to 86.5 ml.
2. Heat to 38 °C with stirring to solubilize the urea, then filter (0.2 μm), and degas.
3. Add solution B: 3.5 ml, with gentle stirring.
4. Firmly tape a 50 cm plastic strip, to be used as a handle, on to the outside of a 100 ml beaker.
5. Pour the IEF gel mixture into the beaker, add 0.1% of solution C, and 0.035% of TEMED.
6. Holding the plastic strip in one hand and the beaker in the other, descend the beaker down into a graduated cylinder (1 litre).
7. Place the IEF glass tubes in the beaker making sure that no air bubbles enter the tubes.
8. Gently layer the deionized water, heated to 40 °C, over the acrylamide to force the IEF gel mixture into the tubes. Stop when the IEF gel mixture reaches a height of 34–35 cm in the glass tubes.
9. Cover the gel surface with 20 μl of deionized water and allow it to polymerize overnight.

B. *Pre-focalization and focalization*

1. Prepare the anodic solution (25 mM phosphoric acid, 3 litres) and the cathodic solution (50 mM sodium hydroxide, 1 litre, degassed).
2. Wash the gel surface with deionized water.
3. Carefully free the glass tubes from the remaining polymerized acrylamide at the bottom of the beaker and remove all of the acrylamide fragments.
4. Block the bottom of the tubes with a piece of dialysis tubing held in place with an elastic band, to avoid the gels sliding out of the tubes.
5. Place the tubes in the running rack.
6. Fill the upper buffer chamber with the cathodic solution until the tube top is under the buffer level.
7. Remove air bubbles on top of the gels with the cathodic solution using a microsyringe.
8. Load 15 μl of GOS on to the top of each gel. Fill the lower buffer chamber with the anodic solution.
9. For pre-focalization: set the maximum voltage to 1800 and maximum intensity to 20 mA. When turning on the main power supply, adjust the voltage to 500 V using the intensity control knob. Pre-focalization is completed when the voltage reaches 1800 (about 90 min).
10. Wash the top of the gels to eliminate the urea.
11. Remove the cathodic solution from the upper chamber, and remove air bubbles on the top of the gel.
12. Load 10 μl of GOS on top of each gel.
13. Carefully load with a microsyringe 100–120 μl of sample, solubilized in sample buffer (solution E), between the GOS and the gel. For best resolution do not load more than 2 mg of protein.
14. Set the maximum voltage to 2000, the maximum intensity to 4 mA, and the maximum power to 7 W.
15. Focalization is finished at 36 000 V h (about 18 h).
16. After electrophoresis, the gels (still in the tubes) can be stored at −20°C, ensuring that the orientation of the gel is clearly marked together with its identity.

Protocol 3. Preparative 2D gels: second dimension

Reagents

- 1.5% agarose solution in 0.125 M Tris–HCl buffer pH 7 containing 1 mM EDTA
- Buffer A: 10 ml of Tris–HCl buffer pH 6.8 (*Protocol 1*, solution B), 3% of SDS, 5 mM of DTT, 0.001% of bromophenol blue, add deionized water up to 100 ml
- Buffer B: 10 ml of solution B (*Protocol 1*), 3% of SDS, 0.2 M of iodoacetamide, add deionized water up to 100 ml

Protocol 3. *Continued*

Method

1. Prepare two glass plate sandwiches (1.5 mm thickness spacers) for each half of the first-dimension gel to be run in the second dimension. Use bevel-edged glass plates. Plates must be turned so as to have a wide space on top of the sandwiches.
2. Prepare acrylamide gels as described in *Protocol 1*, omitting the stacking gel. Fill the glass plate sandwiches to approximately 1 cm from the top with the resolving gel solution and allow the gel to polymerize overnight.
3. Carefully extrude the first-dimension gels pushing them with a 5 ml plastic syringe filled with deionized water.
4. Equilibrate the first-dimension gels (one at a time) successively in 20 ml buffer A and 20 ml buffer B, each for 10 min with stirring.
5. Then put the gel on a piece of Parafilm and cut the gel into two pieces, marking the anodic and cathodic sides on the Parafilm.
6. Pour the agarose solution on to the resolving gel up to the bottom of the bevels to make a good contact between the first-dimension gel and the agarose.
7. Lay each half of the first-dimension gel on to the agarose layer of one second-dimension gel and mark the anodic and cathodic side on the glass plates.
8. Cover the first-dimension half gel with the agarose solution to fix it on to the agarose layer.
9. Fill the anodic chamber of the gel apparatus with electrode buffer as described in *Protocol 1*, and the upper cathodic chamber with two times concentrated electrode buffer, taking care not to detach the first-dimension gel.
10. Set current intensity to 50 mA per gel sandwich.
11. Electrophoresis lasts for about 5 h.
12. Stain the gels as described in *Protocol 4*.

One of the major variables in 2D gel preparations is the composition of ampholytes, and variations are found from batch to batch even from the same source. Since it is frequently necessary to mix different batches of ampholytes having different ranges of pH to optimize the separation of proteins of interest, it is essential that some form of standardization procedure is used when first running 2D gels or when using a new mixture of ampholytes. A very convenient sample is human plasma or serum since numerous laboratories have characterized these protein patterns.

Many things can go wrong when doing 2D gel separations that will lead to non-reproducible results. Most of those technical difficulties can be overcome if the nature of the problem encountered is understood. A troubleshooting section for 2D gels is beyond the scope of this chapter, and excellent discussions of many problems have been already published (for example refs 23 and 24). The 2D gel procedure described in the *Protocols 2* and *3* requires a minimal amount of specialized equipment, but if a large number of gels are to be run on a routine basis it may be helpful to buy specialized equipment to prepare and run several gels at once.

2.3 Staining of the gels

To visualize the proteins after electrophoresis the gels are stained with Coomassie brilliant blue R-250 (CBB). Modifications of the standard CBB staining procedure, which requires fixation in 50% methanol, 10% acetic acid for 2 h, staining with 0.1% CBB in 50% methanol, 10% acetic acid after fixing the gels, and destaining in 40% methanol, 7.5% acetic acid, resulted in the optimized staining procedure described in *Protocol 4*.

Protocol 4. CBB staining of the gels

Reagents

- Solution A (staining solution): 0.1% CBB in 20% methanol, 0.5% acetic acid
- Solution B (destaining solution): 30% methanol

Method

1. Remove the gel assembly from the apparatus, lay it on a flat surface with the cut plate uppermost. Gently prise the two plates apart with a spatula, leaving the gel on the lower plate. Cut one lower corner off the gel to mark the orientation of the samples.
2. Extract the gel from the plate and rinse it in distilled water. Do not fix it.
3. Stain the gel in 300 ml of solution A for 20 min on a shaker platform set at a speed to optimally circulate the solution.
4. Destain in solution B until the protein bands become visible, also on the shaker.
5. Cut with a scalpel the bands of interest, and digest as described in *Protocol 5*.

The CBB stained gels were analysed with the computer-based image analyser system Visage 2000 (Bioimage, Millipore) equipped with the Electrophoresis Quantifier software. Radiolabelled proteins were quantified with the storage phosphor technology. The gels were placed in contact with the storage phosphor screen in a Molecular Dynamics exposure cassette, and the latent image was visualized by scanning the screen in a PhosphorImager (Molecular Dynamics) and analysed with the standard *Imagequant* program.

We observed in preliminary experiments that the yield of in-gel digestion of proteins was improved when the acetic acid and methanol concentrations were reduced in the staining solution of the CBB staining protocol. An optimization of the staining–destaining solutions, based on the digestion yield of known proteins, resulted in the procedure described in *Protocol 4* where, in the CBB staining solution, the acetic acid is reduced from 10% to 0.5% and the methanol from 50% to 20%. In the destaining solution, the acetic acid is eliminated and the methanol is increased from 20% to 30%. The modified procedure was tested on a set of standard proteins separated on a 12.5% gel. As seen in *Table 1*, the gel stained with the modified procedure showed a higher background intensity but, since the intensity of the protein bands was also increased, the sensitivity was not severely affected and approximately 1 μg of the standard proteins were visualized. Furthermore, the amount of protein present in each band after staining–destaining the gels with the two procedures was estimated with a set of radiolabelled standard proteins. The results showed that the amount of the different proteins on the gel was similar in both cases (*Table 1*), indicating that no material was lost when the gel was stained with the modified procedure.

3. Digestion of the proteins in the gel matrix

We observed that the direct digestion of the CBB stained proteins in the gel matrix, after overnight equilibration in ammonium carbonate buffer, gave

Table 1. Staining intensity and recovery of a set of standard radiolabelled proteins stained and destained either with the standard or the modified CBB protocol

Proteins	**Mw**	**IOD**[a]	**dpm**[b]		
	kDa	**CBBs**[c]	**CBBm**[d]	**CBBs**[c]	**CBBm**[d]
Myosine	200	nd[e]	nd	8816	9348
Phosphorylase b	94	3.0	5.9	15 200	13 585
Bovine serum albumin	67	15.5	17.6	8740	9044
Ovalbumin	43	14.5	17.6	13 490	12 141
Carbonic anhydrase	30	12.9	14.6	18 335	15 200
Trypsin inhibitor	20	9.0	9.4	21 090	16 307
α-Lacktalbumin	14	14.2	14.0	nd	nd
Lysozyme	14	nd	nd	8664	12 065
Background[f]	–	2.5	4.0	–	–

[a] Integrated optical density.
[b] Estimated from the sum of the pixel values from the PhosphorImager scanning of the latent image of the gel.
[c] CBB standard staining–destaining procedure.
[d] CBB modified staining–destaining procedure.
[e] Not determined.
[f] Gel region with no protein used to calculate the background IOD.

more complex patterns when compared with the peptide maps obtained for the same protein in solution. Some peaks, present in the fingerprint obtained in solution, are absent, and new ones are detected. Also, an important disturbance of the baseline is observed. Some of the extra peaks are directly related to the CBB dye, but others, as well as the baseline disturbance, are gel-derived impurities that showed only background levels of amino acids after acid hydrolysis. From these observations it was evident that to get better and more reproducible fingerprints, exhaustive washing of the gel slice before digestion was necessary. The method that we developed is described in *Protocol 5*.

Protocol 5. In-gel digestion

Reagents

- Solution A: 50% acetonitrile in 200 mM ammonium carbonate pH 8.9
- Solution B: 200 mM ammonium carbonate pH 8.9, 0.02% Tween 20
- Solution C: 200 μg/ml porcine trypsin in 200 mM ammonium carbonate pH 8.9
- Solution D: 60% acetonitrile, 0.1% trifluoroacetic acid
- 200 mM ammonium carbonate pH 8.9
- Trifluoroacetic acid

Method

1. Wash twice the excised gel pieces with 150 μl of 50% acetonitrile in 200 mM ammonium carbonate pH 8.9, for 20 min at 30°C in an Eppendorf thermomixer.
2. Leave the gel pieces to semi-dry at room temperature (around 10 min depending on the fragment size) on a Parafilm sheet covered with the top of a Petri dish.
3. Partially rehydrate the gel slices with 5 μl of 200 mM ammonium carbonate pH 8.9, 0.02% Tween 20.
4. Add 2 μl of the porcine trypsin solution.
5. After absorption of the protease solution, add aliquots of 5 μl of ammonium carbonate buffer until the gel slices recover their original size.
6. Cut the gel slice into 1 mm^3 pieces, place them in an Eppendorf tube, and add a minimum volume of rehydration buffer to totally immerse the gel pieces (30–50 μl).
7. Carry out the digestions for 4–6 h at 30°C.
8. Stop the digestion by adding 1.5 μl of trifluoroacetic acid.
9. Recover the resulting peptides by three extractions of 20 min each, with 100 μl of a solution D, at 30°C with shaking in an Eppendorf thermomixer.
10. Combine the extracts and concentrate to approx. 20 μl in a SpeedVac (Savant).

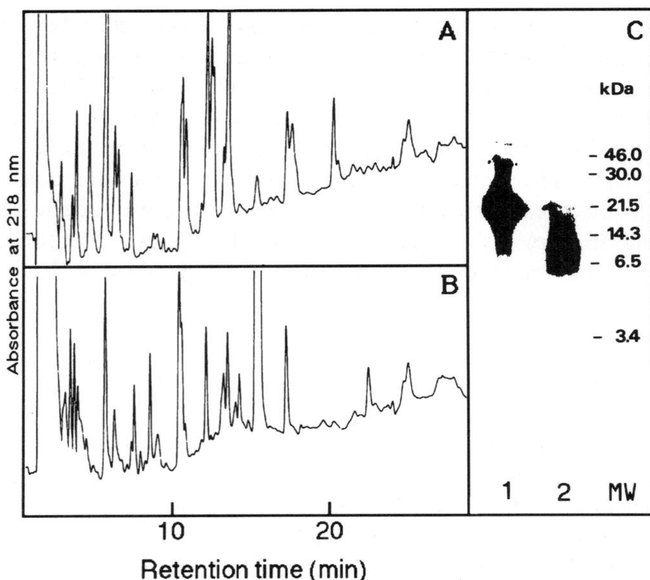

Figure 3. Tryptic digestion of IL-6 (2 μg). (A) Peptide map after digestion in solution. (B) Digestion in-gel as described in Section 3. (C) Computer generated image of the SDS–PAGE exposed to the storage phosphor screen. Lane 1: [^{125}I]IL-6 before digestion. Line 2: SDS–PAGE analysis of the [^{125}I]IL-6 peptides remaining in the gel after in-gel digestion and peptide extraction. The exposure time for line 2 was 20 times that of line. 1.

To wash the gel slices before digestion we retained, after the evaluation of several solvents, a solution of 50% acetonitrile. This washing step is very effective in removing the blue dye and the other contaminants from the gel slices without eluting the protein. *Figure 3A* and *B* shows that the fingerprints obtained from the digestion of 2 μg (approx. 100 pmol) of IL-6 either in solution or in-gel are similar. When the washed gel slices are allowed to partially dry at room temperature they show an important reduction in size, approximately to 20% of the original volume. Then, the gel slices are re-hydrated with a trypsin solution to its original volume. At this step the trypsin efficiently penetrates the gel and incubation for 4 h at 30 °C will result in almost complete digestion, because of the high enzyme to protein ratio (*Figure 3C*). The resulting peptides are recovered by passive elution. Approximately 80% of the material is recovered in the first extraction and the rest in the second. No further peptides are usually detected, by HPLC analysis, in a third extraction. For the in-gel digestion shown in *Figure 3*, the IL-6 was spiked with approx. 70 000 d.p.m. of [^{125}I]IL-6 before running the gel, and the resulting radioactive peptides were eluted from the gel in the two first extractions (approx. 79%, 8%, 2%, and 1% of the total radioactivity respectively for four extractions). The radioactivity that remains in the gel after the peptide extraction (approx. 4100 d.p.m.) corresponds to large peptides of partially digested IL-6, since when analysed in a second SDS–PAGE this radioactivity migrates as discrete bands with a molecular mass greater than 5 kDa (*Figure 3C*) (it should be noted that in *Figure 3C* the exposure time for line 2 is 20 times that of line 1). However, for very hydrophobic proteins the third wash may be necessary. The extracts are combined, concentrated in a SpeedVac, and then the peptides are separated on a narrow-bore HPLC column for analysis by N terminal sequencing and mass spectrometry.

4. HPLC peptide purification and characterization

4.1 Narrow-bore HPLC

The peptides eluted from the polyacrylamide matrix were directly loaded on to a reverse-phase column (C18, 250 mm × 2.1 mm) (Brownlee) and eluted with a linear gradient from 5–60% of acetonitrile containing 0.1% trifluoroacetic acid, in 60 min at a flow rate of 0.3 ml/min using a Brownlee Labs microgradient system. Elution was monitored at 218 nm with a Spectroflow 783 detector. Peaks were manually collected.

The recovery of the peptides after the in-gel digestion is on average 65% when compared, by area, to the peptides obtained from the proteins digested in solution (*Figure 3A* and *B*). These yields were confirmed by amino acid analysis of selected peaks for several different proteins. The sequencing of some of the peptides recovered after in-gel digestion (*Figure 3B*) gave the results shown in *Table 2*.

4.2 Capillary HPLC

The equipment used for capillary HPLC is the same as described in Section 4.1 with the following modifications: a flow reducing system (70 to 1) (LC Packings, San Francisco, CA, USA) was placed at the mixer outlet; all tubing connections downstream of the flow reductor were made using minimum lengths of 0.004 inch i.d. stainless steel tubing. A Delta Pack C18, 5 μm packed capillary column (15 cm × 320 μm) (LC Packings, San Francisco, CA, USA) was connected directly to a Valco Model C14W injector with a 100 ml injector rotor. The outlet of the column was connected directly to a Spectroflow 783 detector equipped with a 2 mm microbore flow cell.

The peptide map of 5 ng of hGH (0.25 pmol) is shown in *Figure 4D*. The chromatographic conditions to produce the shown chromatogram are: flow rate 5 μl/min, gradient 2–60% acetonitrile in 60 min, with detection at 218 nm. The eluted peaks from this run were manually collected on the targets for the mass spectrometric analysis as described and discussed in Section 5.2.

Table 2. Sequence of peptides generated by in-gel digestion of proteins isolated from one- or two-dimensional gels

Sample	Yield[a]	Sequence	Position in the sequence
IL-6 (*Figure 3*)			
Peptide 1	17	NLDAIMXPDPMXNAX	160–175
Peptide 2	28	YILDGISALRXETXNX	59–74
Peptide 3	35	VLIQFLQK	149–156
Peptide 4	4	SFKEFLQSSLX	197–207
Gel 1D (*Figure 5*)			
Spot A (31 kDa) identified as porin[b]			
Peptide 1	45	NINDLLNK	12–19
Peptide 2	52	VSDSGIVXLAYK	237–248
Peptide 3	36	SAVLNTTFTEPFFXAX	109–124
Gel 2D (*Figure 2*)			
Spot A (65 kDa, pI 5.3) identified as heat shock cognate protein[b]			
Peptide 1	37	NSLESYAFNMK	540–550
Peptide 2	24	SFYPEEVSXMVLTX	113–126
Peptide 3	18	TVTNAVVTVPAYFNDXQX	138–155
Spot B (43 kDa, pI 7.3) identified as phosphoglycerate kinase[b]			
Peptide 1	22	YSLEPVAAELK	146–150
Peptide 2	31	VLNNMEIGTSLYDEEGAX	246–263

[a] Initial yield calculated by extrapolation of the PTH–amino acid values to cycle 1.
[b] The proteins were identified by comparison with the sequences present in the SwissProt database.

5. Peptide characterization and protein identification

5.1 Peptide sequencing and amino acid analysis

The peptides isolated by narrow-bore HPLC as previously described can be directly sequenced. In our laboratory the peptide solution is spotted on to a polybrene pre-treated glass fibre filter and sequenced in a gas phase protein sequencer (470A, Applied Biosystems). Whenever possible, an aliquot of the purified peptide is used for amino acid analysis in a 420A amino acid analyser, with an on-line 130A PTC analyser of Applied Biosystems. Amino acid compositions are very useful to estimate recoveries and thus follow the efficiency of the overall procedure.

In general, for in-gel digestions of proteins detected with CBB staining, the initial yields for at least some of the isolated peptides should be sufficiently high, 10 pmol or more, to allow the identification of long stretches of amino acids with very few ambiguities (*Table 2*). We have noted that often we are able to identify the PTH–amino acids of labile residues like methionine and

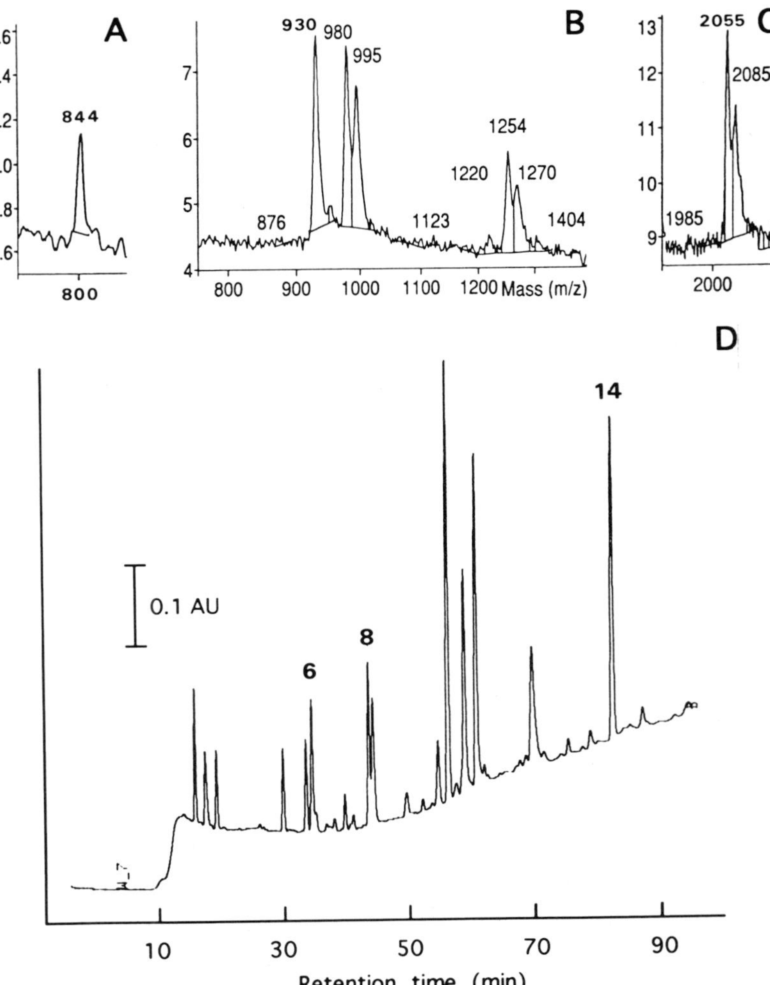

Figure 4. Capillary HPLC of hGH. 5 ng of a tryptic digestion of hGH were directly loaded onto the capillary HPLC column and the eluted peptides were collected for mass spectrometric analysis as described in Section 5.2. The mass spectra of peaks 6, 8, and 14 are shown in inset A, B, and C.

tryptophan. These results are probably the consequence of the elimination of the electrotransfer step, the short time needed to obtain the peptides for sequencing, and to the overnight polymerization of the gels. PTH–serine were also recovered quite efficiently when the peptides were sequenced immediately after purification. If not possible, the sample should be stored at −20 °C.

5.2 Mass spectrometry

Analysis by mass spectrometry (MS) of the peptides eluted from the polyacrylamide matrix were performed on a laser desorption time of flight mass spectrometer. Lasermat (Finnigan MAT Ltd, UK) either directly after the SpeedVac concentration, or on the purified peptide solution used for sequence and amino acid analysis. When very little amount of material is available we purify the peptides by capillary HPLC and use them only for MS analysis. For the direct MS analysis of the peptide mixture, or after narrow-bore HPLC purification, the peptide solution, 0.2 μl is mixed with 0.5 μl of a solution of α-cyano-4 hydroxycinnamic acid (5 mg/ml in 70% acetonitrile, 0.1% trifluoroacetic acid) on the target of the mass spectrometer and allowed to dry at room temperature. For the MS analysis after capillary HPLC, each peak is directly manually collected on a target of the mass spectrometer on which 0.5 μl of the solution of α-cyano-4 hydroxycinnamic acid has been previously added and dried, then another 0.5 μl of the solution of α-cyano-4 hydroxycinnamic acid is added and dried at room temperature before the analysis. The peptide mass values for each digestion sample are the average of at least four analysis, and each analysis consists of at least 20 laser pulses. We choose the matrix assisted laser desorption time of flight mass spectrometer (LDMS) technique because of its high sensitivity, its capacity for recording mass spectra of complex mixtures in the mass range of the resulting tryptic peptides, its simplicity, and the short time required for analysis of individual samples (25). Furthermore, we speculate that since the SDS, an inhibitor for the matrix assisted LDMS (26), is eliminated during the in-gel digestion (18), the tryptic digests may be analysed directly after extraction from the gel matrix, thereby simplifying the sample preparation and reducing the analysis time. In fact, the analysis of the in-gel tryptic digestion of the hGH produces results similar to those obtained when the protein was analysed after digestion in solution. And the high sensitivity of this approach can be seen by the production of MS results from subpicomol quantities of digested hGH. *Figure 4D* shows the capillary HPLC of 0.25 pmol of hGH digested with trypsin; the eluting peptides were collected as described and MS analysis allowed their identification. For example, *Figure 4A* allows the identification of peak 6 as the peptide hGH 71–77 844 mass units (m.u.). *Figure 4B* shows that three peptides coelute: hGH 1–8 (930 m.u.), hGH 9–16 (980 m.u.), and hGH 169–178 (1254 m.u.). It should be noted that the peaks at 995 and 1270 m.u. may indicate oxidation of the methionine residues present in peptides hGH 9–16 and hGH 169–178. *Figure 4C* identifies peak 14 as peptide hGH 78–84 (2055 m.u.); in this case the second peak at 2085 m.u. may represent oxidation of a tryptophan residue present in this peptide.

5.3 Database search with the peptide mass fingerprints

The experimental mass values of the peptides, together with the pI and the M_r estimated from the position of the spot in the 2D gel, were used to search the protein sequence databases to ascertain whether the protein was already known. To this end a program that computes the theoretical pI and M_r values for each database entry was used. This program is similar to those already described (27). Our program (28) first selects all proteins having pI and M_r within $\pm$ 15% of the corresponding M_r and pI values calculated from the 2D gel for a particular spot (for proteins isolated in 1D gels the pI values are set to 1–14). Then, for each selected protein the theoretical M_rs of the peptides that would result from a tryptic digestion were calculated. Finally, the theoretical mass values of the tryptic peptides were compared with the experimental values obtained from the protein digested in the gel matrix. The speed of the searching process was increased by using a database that we built and that included all the theoretical peptide M_rs together with the protein pI and M_r information.

6. Examples of the identification of proteins after in-gel digestion

The described methods are illustrated by the identification, through internal sequence analysis, of yeast mitochondrial proteins separated on SDS–PAGE, and of rat brain proteins separated on preparative 2D gels.

The mitochondrial proteins from yeast, 50 μg, were directly separated on a 12.5% SDS–PAGE (*Figure 5*). The indicated band was cut out and in-gel digested with trypsin. The extracted peptides from the digestion of the 32 kDa band gave the fingerprint shown in *Figure 5B*. Sequencing of several peptides (*Table 2*) resulted in the identification of this protein as porin, an outer mitochondrial membrane protein. *Figure 5A* shows the results obtained by MS analysis of the peptides extracted from the gel after digestion. A database search with the MS data showed that eight out of ten values corresponded to tryptic peptides of the yeast porin (theoretical values: 1202.4, 1342.5, 1474.6, 1802.0, 1958.3, 2061.2, 2268.4, and 3258.5 m.u.). These results are in line with the sequence data and validates the assignment of the band as porin.

Figure 2 shows a representative CBB stained preparative 2D gel. The first half corresponds to the pI region 4.0–5.7 and the second to the pI region 5.7–7.4. The total protein loaded in this gel was 1.7 mg, more than 200 protein spots were detected, and almost all of these spots could be directly in-gel digested, and either characterized by MS analysis or sequenced. As an example, we show the MS analysis and the peptide maps resulting from the in-gel digestion of two of those spots. The spot A (65 kDa, pI 5.3) was subjected to in-gel digestion, and the resulting peptide map is shown in *Figure 6B*. The indicated peptides were sequenced (*Table 2*). The sequences obtained were correlated with the heat shock cognate 70 kDa protein, a protein present in the database. *Figure 6A* shows the MS analysis of the peptides extracted from the gel after digestion. The computer program developed to search in the database found that six out of ten of the values correspond to tryptic peptides of this

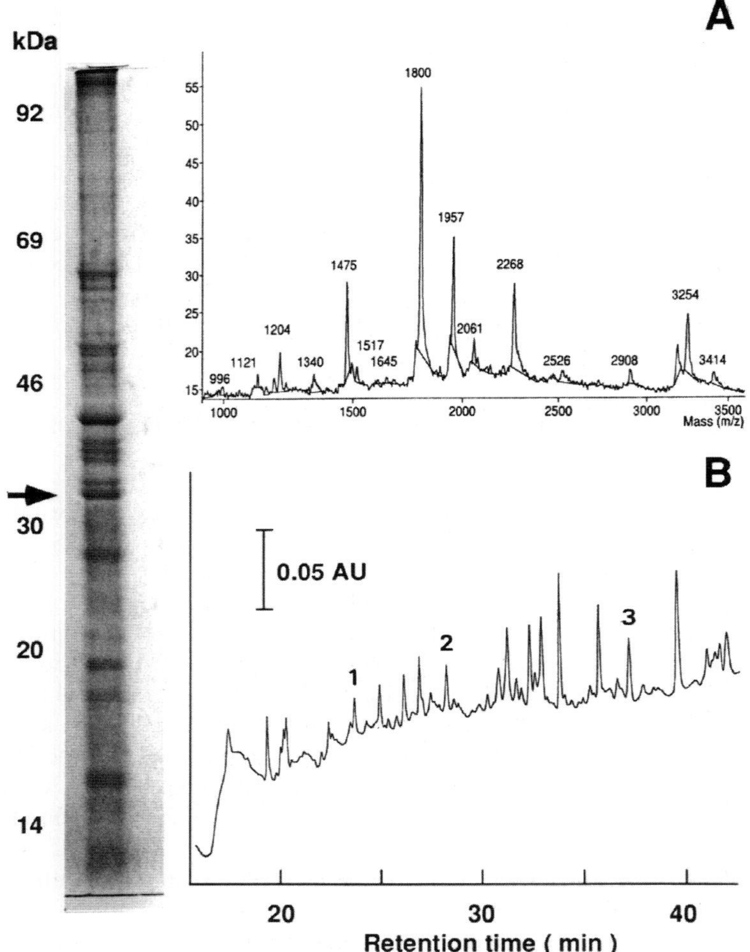

Figure 5. Gel electrophoresis of yeast mitochondrial membrane proteins. 50 μg of protein were separated on a 12% polyacrylamide gel, stained as described in Section 2.3, and the indicated band was in-gel digested. (A) Mass analysis of the eluted peptides. (B) Narrow-bore HPLC separation of the peptides. The peptides 1–3 were sequenced and the results are shown in *Table 2*.

heat shock protein (theoretical values: 1254.4, 1481.7, 1691.7, 1982.2, 2775.0, and 2998.2 m.u.) confirming the sequence data and thus, the identification of the protein.

The spot B (43 kDa, pI 7.5) gave the fingerprint shown in *Figure 7B*. The sequences of three peptides (*Table 2*) allowed us to identify this spot as phosphoglycerate kinase. The molecular weight and pI calculated from the

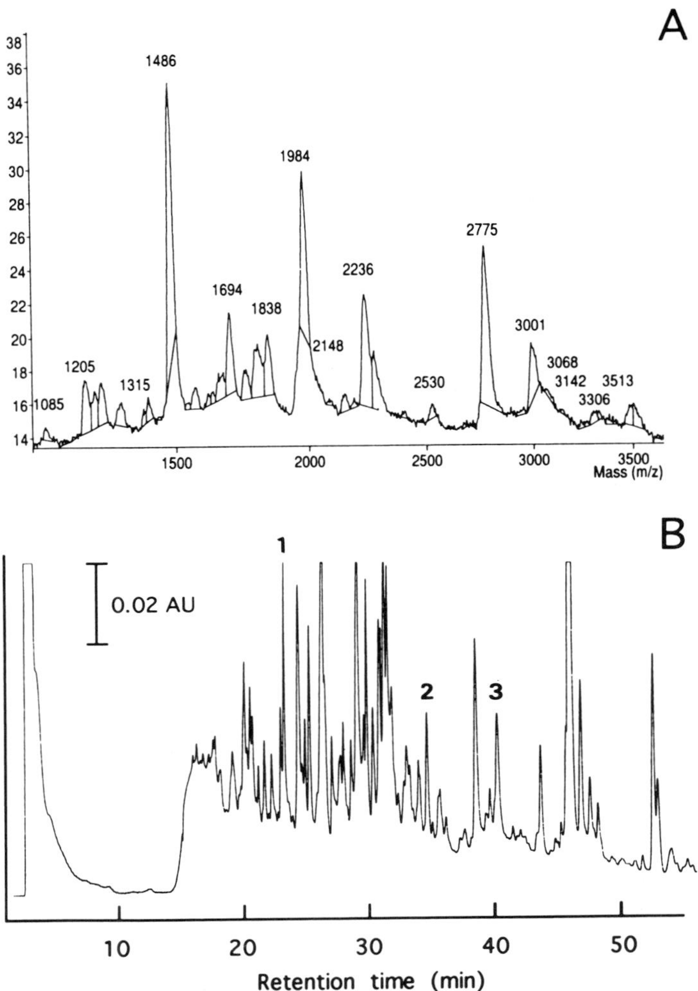

Figure 6. Analysis of spot A isolated by preparative 2D gel electroophoresis (*Figure 2*). The eluted peptides resulting from the tryptic digestion of spot A were analysed by mass spectrometry (A), and then separated on narrow-bore HPLC (B). The indicated peptides 1–3 were sequenced and the results are shown in *Table 2*.

sequence are in agreement with the value determined from the gel. Furthermore the MS analysis (*Figure 7A*) showed that six out of nine of the values correspond to tryptic peptides of phosphoglycerate kinase (theoretical values: 1219.4, 1634.8, 1769.0, 1999.2, 2105.4, 2511.7 m.u.), thus validating the sequence data and showing again that the values obtained from the MS analysis can be very helpful for the identification of proteins after in-gel digestion.

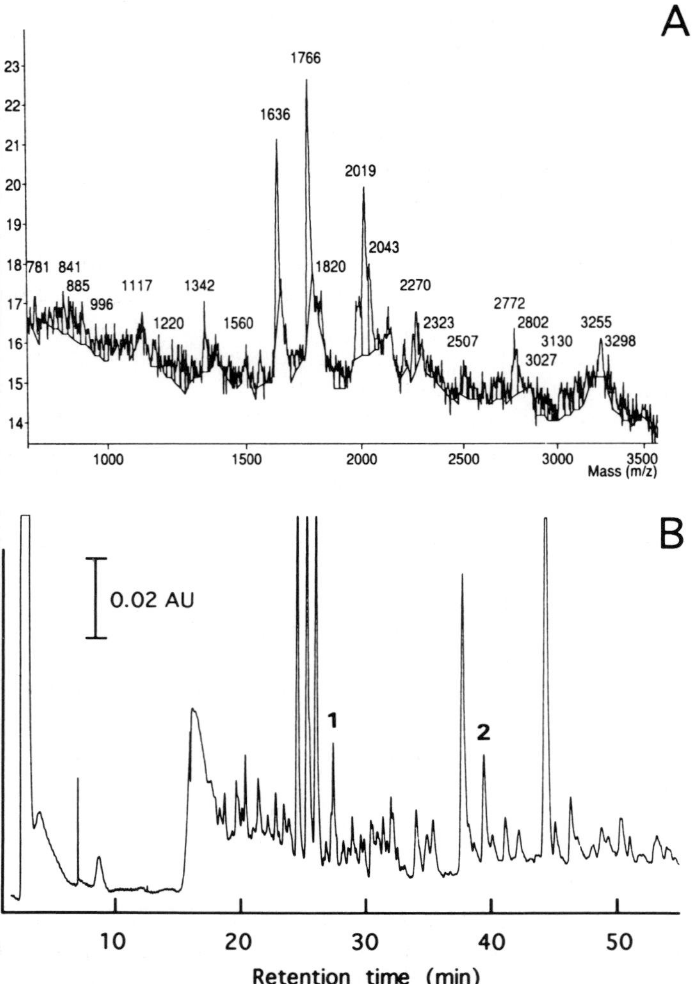

Figure 7. Analysis of spot B isolated by preparative 2D gel electrophoresis (*Figure 2*). The eluted peptides resulting from the tryptic digestion of spot B were analysed by mass spectrometry (A), and then separated on narrow-bored HPLC (B). The indicated peptides 1 and 2 were sequenced and the results are shown in *Table 2*.

7. Conclusions

The production of sequence information from proteins isolated by gel electrophoresis is a feasible approach to their chemical characterization. Moreover, if one can obtain internal peptides by an in-gel digestion technique the problem of natural or artefactual N terminal protein blockage is circumvented.

As shown in some of the examples described here, we succeeded in sequencing membrane proteins of the mitochondria, and therefore the method should be of particular utility for sequencing hydrophobic insoluble membrane proteins difficult to isolate by other means.

It should be noted too that the coupling of the resolving power of 2D gels with the high sensitivity of matrix assisted LDMS analysis of peptides after in-gel digestion has several practical consequences. Peptide mass data can be collected and stored in 2D gel databases from a large number of spots in an extremely short time. Furthermore, since the amount of material used is very small, the approach can be easily used as a screening procedure before sequencing peptides from in-gel digested proteins. The peptide mass data, together with the M_r and pI of the proteins calculated from the position in the gels, can be used to identify the proteins even if the amount of material present in the gel is below the level necessary for microsequencing. The peptide mass data can also help to correlate new protein sequences derived from cDNA sequencing, with spots present in 2D gels. Additionally, the peptide mass data, even for unidentified proteins, may be useful for the comparison of spots from different gels, thereby allowing the standardization of the different databases that are being constructed. Matrix assisted LDMS analysis has been recently used to produce sequence information from peptides (29). The coupling of the procedures described here with that technology should facilitate the characterization of minor components in complex mixtures.

In summary the described methods open a new possibility to identify proteins isolated by electrophoresis techniques and also, for example, to link 2D gels with protein and DNA databases, that may facilitate the biological interpretation of 2D gel analysis.

Acknowledgements

I thank Drs D. Boureme, J. C. Guillemot, J. Rosenfeld, A. Pradines, E. Ferran, and J. Capdevielle for all the work and discussions that made possible all the methods and results described. I also thank Drs D. Caput and N. Vita and D. Shire for stimulating discussions and comments on the manuscript.

References

1. Latham, K. E., Beddington, R. S., Solter, D., and Garrels, J. I. (1993). *Mol. Reprod. Dev.*, **35**, 140.
2. Baker, C. S., Corbett, J. M., May, A. J., Yacoub, M. H., and Dunn, M.J. (1992). *Electrophoresis*, **13**, 723.
3. Jungblut, P., Dzionara, M., Klose, J., and Wittmann-Leibold, B. (1992). *J. Protein Chem.*, **11**, 603.
4. Celis, J. E., Rasmussen, H. H., Leffers, H., Madsen, P., Honore, B., Gesser, B., *et al.* (1991). *FASEB J.*, **5**, 2200.

5. Aebersold, R. and Leavitt, J. (1990). *Electrophoresis*, **11**, 517.
6. Matsudaira, P. (1987). *J. Biol. Chem.*, **262**, 10035.
7. Moos, M. Jr., Nguyen, N. Y., and Liu, T. Y. (1988). *J. Biol. Chem.*, **263**, 6005.
8. Aebersold, R. H., Leavitt, J., Saavedra, R. A., Hood, L. E., and Kent, S. B. (1987). *Proc. Natl. Acad. Sci. USA*, **84**, 6970.
9. Tempst, P., Link, A. J., Riviere, L. R., Fleming, M., and Elicone, C. (1990). *Electrophoresis*, **11**, 537.
10. Simpson, R. J., Ward, L. D., Reid, G. E., Batterham, M. P., and Moritz, R. L. (1989). *J. Chromatogr.*, **476**, 345.
11. Szewczyk, B. and Summers, D. F. (1988). *Anal. Biochem.*, **168**, 48.
12. Xu, Q. Y. and Shively, J. E. (1988). *Anal. Biochem.*, **170**, 19.
13. Hager, D. A. and Burgess, R. R. (1980). *Anal. Biochem.*, **109**, 76.
14. Hunkapiller, M. W., Lujan, E., Ostrander, F., and Hood, L. E. (1983). In *Methods in enzymology* (ed. C. H. W. Hirs and S. N. Timasheff), Vol. 91, pp. 227–36. Academic Press, New York.
15. Wessel, D. and Flugge, U. I. (1984). *Anal. Biochem.*, **138**, 141.
16. Simpson, R. J., Moritz, R. L., Nice, E. C., and Grego, B. (1987). *Anal. Biochem.*, **165**, 21.
17. Ward, L. D., Reid, G. E., Moritz, R. L., and Simpson, R. J. (1990). *J. Chromatogr.*, **519**, 199.
18. Rosenfeld, J., Capdevielle, J., Guillemot, J. C., and Ferrara, P. (1992). *Anal. Biochem.*, **203**, 173.
19. Ferrara, P., Rosenfeld, J., Guillemot, J. C., and Capdevielle, J. (1993). In *Techniques in protein chemistry IV* (ed. R. H. Angeletti), pp. 379–87. Academic Press Inc., San Diego.
20. Laemmli, U. K. (1970). *Nature (London)*, **227**, 680.
21. O'Farrel, P. (1975). *J. Biol. Chem.*, **250**, 4007.
22. Anderson, N. G. and Anderson, N. L. (1978). *Anal. Biochem.*, **85**, 331.
23. Dunbar, B. S. (ed.) (1987). *Two-dimensional electrophoresis and immunological techniques*. Plenum Press, New York.
24. Sinclair, J. and Rickwood, D. (ed.) (1981). *Gel electrophoresis of proteins: a practical approach*. IRL Press, Oxford.
25. Chait, B. T. and Kent, S. B. H. (1992). *Science*, **257**, 1885.
26. Mock, K. K., Sutton, C. W., and Cottrell, J. S. (1992). *Rapid Commun. Mass Spectrom.*, **6**, 233.
27. Pappin, D. J. C., Hojrup, P., and Bleasby, A. J. (1993). *Curr. Biol.*, **3**, 327.
28. Sagliocco, F., Guillemot, J.-C., Monribot, C., Capdevielle, J., Perrot, M., Ferran, E., Ferrara, P., and Boucherie, H. (1996). *Yeast*, **12**, 1519–33.
29. Chait, B. T., Wang, R., Beavis, R. C., and Kent, B. H. (1993). *Science*, **262**, 89.

3

Assembly of MHC molecules: requirements of glycosylation, oxidized glutathione, and peptides

RANDALL K. RIBAUDO

1. Introduction

The ordered assembly of multisubunit proteins such as MHC molecules from their component chains is a fundamental process in the generation of functional complexes such as receptors, signalling complexes, and presentation molecules. Typically, these processes involve not only the constituent chains which make up the mature complex, but also rely on the action of chaperones, and other accessory proteins. Assembly of the two chains of class I MHC molecules occurs in the endoplasmic reticulum and is dependent on the simultaneous presence of the polymorphic heavy chain, and the non-polymorphic light chain, β2-microglobulin. Careful pulse chase analysis has demonstrated that the two chains rapidly associate in the endoplasmic reticulum (1), and remain there associated with the molecular chaperone calnexin until they acquire an appropriate peptide (2–5). At this point the trimeric complex can exit the ER and traverse the golgi to the cell surface. From the time of initial translocation of the independent chains to the ER to their final appearance on the cell surface, domains fold, disulfide bonds form, carbohydrate residues are added, and peptide is acquired.

Much of the information regarding this assembly of class I molecules has been obtained by studies using whole cells, with and without defects in various points in the assembly pathway. While this type of approach has generated valuable information about the synthesis, folding, assembly, and transport of these molecules, other *in vitro* systems have been developed that can substantially supplement studies in whole cells to further characterize class I MHC. One such system relies on the *in vitro* translation and assembly of class I MHC proteins from *in vitro* transcribed RNAs which code for class I heavy chains and β2-microglobulin. In this system the only proteins translated and therefore radiolabelled (as a result of incorporation of radiolabelled amino acids) are those encoded by the input RNA, which facilitates subsequent analysis of the products. As this is a cell-free system which contains cellular components required for translation and translocation only as far as the endoplasmic reticulum (there are no golgi membranes present) additional complexities and protein modifications distal to those occurring in the ER can be dissected away from earlier folding and assembly events, allowing careful study in this critical compartment. Additionally, advances in the field of molecular biology have simplified the generation of deletions, substitutions, chimeric constructs, and other modifications to the cDNAs which can then be rapidly transcribed, translated, and evaluated. The refinement and utilization of *in vitro* translation assays to study protein folding and chain association has greatly facilitated our understanding of the underlying mechanisms and requirements for class I assembly. These assays were originally employed in attempts to examine class I MHC assembly as long ago as 1979 (6). In those studies, total mRNA from $H2^d$ lymphoma cells was translated *in vitro* using rabbit reticulocyte lysate and the MHC class I molecule $H2\text{-}D^d$ was immunoprecipitated using both polyclonal, and monoclonal antibodies. Translation, translocation, core glycosylation, and other post-translational modifications could be demonstrated. However, actual assembly in these early studies was not demonstrated, as conditions necessary to allow native folding and disulfide bond formation to occur were not yet known (discussed below). More recently a number of laboratories including our own have reported the *in vitro* translation, translocation, and assembly of class I molecules, using homogeneous RNAs coding for specific MHC chains of interest (7–11). These studies have taken advantage of the flexibility of current *in vitro* translation systems, and the use of homogeneous input RNAs; using them to define more precisely the requirements necessary to allow the native folding and assembly of the newly synthesized chains, such as the requirement for an appropriate redox potential to facilitate disulfide bond formation (10, 11). In this chapter, I will provide a practical detailed description of how to set-up and use an *in vitro* translation system, commenting on parameters which can be varied for specific purposes, as they affect the folding, disulfide bond formation, and assembly of newly translated class I proteins. Detailed discussion of some of the variables should help the investigator customize this system for his/her own purposes.

Parameters of the rabbit reticulocyte lysate based on *in vitro* translation and processing systems can be individually controlled to allow more precise evaluation of folding and assembly. Commercially prepared *in vitro* translation kits are available as core systems consisting of micrococcal nuclease treated lysate, to which creatine phosphate and creatine phosphokinase has been added as a source of ATP. To this, varying amounts of other essential components can be added as desired to optimize conditions depending on experimental requirements. These components which will be individually discussed below include dithiothreitol (DTT) (effects redox potential, and therefore disulfide bond formation), K^+ and Mg^{2+} (important for optimizing translation of RNAs from different sources), and microsomes (necessary for post-translational modifications, such as signal peptide cleavage and core glycosylation). Following translation, the microsomes which contain the newly synthesized proteins can be readily isolated, and products evaluated, either for conformational changes, assembly, or ability to bind peptides, using a variety of analytical techniques. Following is the basic protocol, and a discussion of the various parameters that can be manipulated, how they affect *in vitro* translation and assembly, and some suggestions from our own experience.

2. General protocol

2.1 Reagents

Rabbit reticulocyte lysate: micrococcal nuclease treated. The lysate is prepared from the blood of phenylhydrazine treated rabbits essentially by the original method of Pelham and Jackson (12). The lysate is treated with micrococcal nuclease to destroy endogenous RNA (which is largely globin mRNA). As this nuclease is exquisitely calcium-dependent, the treated lysate is then chelated with EGTA to inactivate the nuclease (see discussion of calcium below). Nuclease treated rabbit reticulocyte lysate is commercially available as a kit from a number of suppliers, which in addition to nuclease treatment, has been supplemented with:

- ATP generating system (creatine phosphate and creatine phosphokinase)
- DTT
- tRNA (from calf liver: expands the range of mRNAs that can be translated)
- haemin (prevents inhibition of initiation factor eIF-2A)

To this lysate is typically added:

- amino acid mixture (generally deficient in one amino acid to enhance incorporation of a radiolabelled derivative)
- RNasin: a placental RNase inhibitor
- magnesium acetate (Mg^2+ is essential for translation—see below)
- potassium chloride (K+ concentration has a dramatic effect on translation efficiency—see below)

Note: kits are also available in which some of the above components (DTT, KCl, MgOAc) are supplied separately to provide greater flexibility (e.g. Promega's 'Flexi-Rabbit Reticulocyte Lysate System').

Additionally, the following reagents are needed:

(a) Microsomes: allows post-translational processing such as signal peptide cleavage, core glycosylation, and disulfide bond formation. Typically, canine pancreatic microsomes are used because of their high processing activity and commercial availability, but microsomes from many sources have been used (discussed below).

(b) Oxidized glutathione (GSSG): provides a redox potential that will support disulfide bond formation.

(c) [^{35}S]methionine: *in vivo* labelling grade, > 1200 Ci/mmole (e.g. Amersham SJ1015).

(d) RNA: *in vitro* transcribed—see below.

(e) Microsome isolation buffer (MIB): 0.75 M KCl, 20 mM Tris–HCl pH 7.5, 10 mM EDTA.

(f) 0.5 M sucrose cushion: 0.5 M sucrose, 20 mM Tris–HCl pH 7,5, 10 mM EDTA.

(g) Lysis buffer (LB): 150 mM NaCl, 10 mM Tris–HCl pH 7.5, 1.5 mM $MgCl_2$, 1% NP-40, 0.03 M iodoacetamide, 1% aprotinin, 0.5 mM AEBSF, 10 mg/ml leupeptin. Anti-proteases aprotinin, AEBSF, and leupeptin may be supplemented, replaced, or omitted, as particular needs require. NP-40 is added to solubilize the microsomal membranes, while not disrupting assembled class I complexes. However, other detergents can be substituted as required.

3. General translation and processing protocol

The protocol listed below is for a typical 25 μl reaction using Promega Corporations Rabbit Reticulocyte Lysate system supplemented with their canine pancreatic microsomes. This reaction can be scaled up or down as needed.

3.1 Preparation of RNA

There are many commercial systems available for the preparation of RNA. Generally, cDNAs can be cloned into a plasmid that has a transcriptional promoter, such as the bacterial promoters SP6, T7, or T3. The plasmid is linearized 3′ of the cDNA, and transcribed *in vitro*. Two developments in this area have improved the quality of RNA and its subsequent translation *in vitro*. One is the development of large scale transcription systems (e.g. Ribomax System, Promega Corporation), which allow the transcription of milligram quantities of RNA that does not inhibit translation in rabbit reticulocyte lysate translation systems even at high concentrations which was a limitation of older transcriptional systems. This improvement is attributed in part to the use of a different buffer system (Hepes versus Tris–HCl), and the presence of inorganic pyrophosphatase. It should be noted that this system was designed and optimized for large scale translation of the resulting RNAs, while older traditional methods were typically used to generate labelled RNA probes for use in Northern and Southern blotting.

The second advancement is the use of the 5′ non-coding region from the encephalomyocarditis virus (EMCV) cloned immediately upstream from the coding region of the cDNA of interest which is discussed below.

3.2 Pre-treatment of RNA

It is generally recommended to heat the RNA to 67°C for 10 min and immediately place on just ice prior to addition to the translation reaction. This reduces secondary structure of the RNA especially in GC-rich regions which can inhibit translation efficiency.

Protocol 1. Translation of RNA

Equipment and reagents

- See Section 2.1
- Microcentrifuge
- Water-bath

Method

1. Combine the following components in order and on ice in an RNase-free microcentrifuge tube:
   - rabbit reticulocyte lysate 17.5 μl
   - amino acid mixture (1 mM) 0.5 μl
   - RNasin 0.5 μl
   - [^{35}S]methionine 2.0 μl
   - canine pancreatic microsomes 1.8 μl
   - 0.1 M GSSG 0.5 μl
   - RNase-free dH_2O q.s. to 24.0 μl
   - RNA 1.0 μl
2. Briefly mix the components by gentle vortexing, collect by centrifugation (5 sec in microcentrifuge), and incubate for 60–90 min at 26–30°C
3. Following translation, stop the reaction by placing on ice.

Protocol 2. Isolation of microsomes

Equipment and reagents

- See Section 2.1
- Beckman Airfuge and rotor

Method

1. Add 2 vol. microsome isolation buffer (MIB) and layer on to a 0.5 M sucrose cushion.
2. Isolate the microsomes by centrifugation at 100 000 *g* for 15 min at 4°C. This is conveniently done in a Beckman Airfuge, using an A100/18 rotor at 22 psi. The microsomes will pellet through the sucrose, leaving the lysate in the upper layer.
3. Aspirate off the lysate-containing supernatant leaving the transparent microsomal pellet.

Assembly of the chains can be evaluated by immunoprecipitation of the translation products following solubilization of the microsomal pellet. Typically, this is done by resuspending the microsomal pellet in an isotonic NP-40 lysis buffer (LB). The buffer contains anti-proteases leupeptin, ABESF (4-(2-aminoethyl)-benzenesulfonyl fluoride), and aprotinin, as well as iodoacetamide, which alkylates free sulfhydryl groups thus preventing artefactual disulfide bond formation. To this buffer, peptides can be added to determine their effects on class I assembly. Typically, the presence of an appropriately MHC restricted peptide at a concentration of 20 μM in the lysis buffer will induce the post-translational assembly of up to 60% of the microsome-associated chains as determined both by generation of a native α1α2 domain, and by the co-precipitation of β2-microglobulin with monoclonal anti-heavy chain antibodies (10), but depends on the binding affinity of the particular peptide to the MHC molecule being studied. Numerous descriptions of peptide translocations into microsomes using three types of *in vitro* translation reactions have been reported in efforts to characterize the ATP-dependence of these reactions (8). However, careful experimentation has demonstrated that there is an ATP-dependent transport of peptides into microsomes, which can be masked by ATP-independent peptide binding to the integral membrane transporters which remain on the outer surface of the membranes until lysis (13). Thus careful considerations of incubation time, temperature, and conditions of membrane lysis must be considered when attempting to study peptide translocation using this system.

3.3 Variable parameters

3.3.1 DTT levels

The formation of disulfide in transmembrane and secreted proteins occurs in the endoplasmic reticulum, and is dependent on a favourable redox potential. As disulfide bonds form, a resident ER enzyme, protein disulfide isomerase (PDI) facilitates the shuffling of these disulfide bonds in order to achieve a native folded state (14–16). *In vivo*, the cytoplasm has a strong reducing environment, which likely promotes translocation of nascent polypeptides into the ER by inhibiting the formation of secondary structure. Within the ER, the redox potential is more oxidizing as a consequence of the presence of oxidized glutathione (GSSG) (17). Most commercially available rabbit reticulocyte lysate and microsome preparations have DTT added (typically 2 mM), and as such will not support disulfide bond formation. However addition of GSSG to the translation reaction, to a final concentration of between 2-4 mM allows the formation of disulfide bonds (10, 11, 18). Class I MHC heavy chains contain two disulfide bonds, between C101 and C164, and between C203 and C259. Additionally, β2-m has one disulfide bond, between residues C25 and C80. By titrating the amount of added GSSG, sequential disulfide bond formation of class I MHC heavy chains can be visualized by analysis using non-reducing

SDS–PAGE (10). Generation of the disulfide bond in the $\alpha 3$ domain of the class I heavy chain appears to be necessary and sufficient to generate the native conformation of this domain as determined by reactivity with monoclonal antibodies (10). In practice, too high a concentration of GSSG will inhibit the translation/translocation reaction, so only enough to compensate for the levels of DTT should be employed. Addition of GSSG does not appear to be necessary for the formation of disulfide bonds when using *in vitro* translation kits which do not have DTT added such a Promega Corporation's Flexi-Lysate system (R. K. R. unpublished observations).

3.3.2 K^+ and Mg^{2+} levels

RNAs from different sources have different optimal salt requirements. Some of these requirements are somewhat generalizable, while others must be empirically determined to obtain optimal translation. Mg^{2+} is absolutely required for translation *in vitro* and the range of Mg^{2+} for optimal translation is very narrow. Therefore, small changes in Mg^{2+} concentration can dramatically affect the efficiency of translation. Most rabbit reticulocyte lysate contains Mg^{2+} at about 1.8–2.5 mM. The need for additional Mg^{2+} must be determined empirically, but generally does not exceed an additional 2 mM concentration. Optimal K^+ concentrations can vary much more and also depend on the source of RNA. Thus, RNAs containing their endogenous 5′ untranslated region, and a reasonably good Kozak sequence (19) with no poly(A) tail translate most efficiently when K^+ concentrations are between 35–45 mM. With the addition of a 30 base poly(A) tail, the optimal K^+ concentration rises to about 65–75 mM. For constructs of this type, the K^+ can be supplied either as KCl, or KOAc, with KCl having a sharper optimal dose-dependence.

3.3.3 Capping of RNA

The addition of a cap structure to the 5′ end of RNA as occurs normally in eukaryotic cells, has been reported to stabilize the RNA. Additionally, translation of some RNA is significantly more efficient if the RNA is capped. However there are also many RNAs for which capping does not appear to be necessary, and therefore this must also be determined empirically. RNA can be capped co-transcriptionally by replacing the rGTP with a 10:1 mixture of $m^7G(5')ppp(5')G$:rGTP in the transcription reaction keeping the total rGTP plus GTP analogue concentration constant. Consequently, approximately 90% of the *in vitro* transcribed RNA will begin with the capped rGTP. Addition of the 7-methyl cap typically results in a decrease in the total amount of RNA transcribed (typically 20–30% lower) but ensures that at least 90% will be capped.

3.3.4 EMCV 5′UT

An alternative to capping RNA is to use transcription vectors which utilize the 5′ untranslated region from the encephalomyocarditis virus (EMCV) such as is used in the pCITE vector system (Novagen). This 5′UT, which replaces the endogenous 5′UT acts as a cap-independent ribosomal binding site, thus obviating the need for capping the *in vitro* transcribed RNA. Studies have demonstrated that optimal translation efficiency with EMCV RNA requires higher K^+ concentrations than observed with their endogenous 5′UT, with the optimal concentrations between 90–115 mM, and generates as much as sevenfold more translation product than comparable RNA containing its own 5′UT (20). Additionally, these studies reveal more efficient and faithful translation initiation when K^+ was supplied as KCl, as opposed to KOAc.

3.3.5 Microsomes

A very powerful aspect of the *in vitro* translation and assembly system is the ability to use microsomes from different cell sources. This can give tremendous flexibility to the study of class I assembly. Biologically active microsome preparations from a variety of cell types have been used and include in addition to canine pancreas, Raji, CHO, and T1 cells, T2 cells which have a defect in the ATP-dependent peptide transport machinery, and Daudi cells which do not synthesize β2-m (8, 9, 21–23). The use of microsomes from cells with specific defects could be a powerful tool in determining the role of various ER processes and proteins in controlling the assembly of class I MHC complexes.

3.3.6 Calcium

Commercially available rabbit reticulocyte lysates are typically treated to remove endogenous RNAs. This is generally done by the addition of micrococcal nuclease to the lysate preparation. This nuclease is exquisitely calcium-dependent, and as such is completely inactivated by the addition of EGTA. As translation is not calcium-dependent, this allows for the simple removal of RNA (and high background translation activity). Following nuclease treatment, the enzyme is inactivated by addition of EGTA. Consequently, when using nuclease treated lysate care must be taken to avoid the addition of calcium to the translation reaction as this would result in reactivation of the enzyme. However, this property can be exploited to allow the sequential translation of proteins in order to examine the kinetic order of assembly. RNA encoding one protein can be translated, followed by addition of calcium in order to destroy the template, followed by EGTA treatment to once again inactivate the enzyme, followed by a second round of translation.

4. General considerations for studying class I MHC assembly

Three primary limitations to the study of class I MHC assembly are the availability of cDNAs, monoclonal antibodies to different domains of the class

I molecule, and appropriate peptides that will bind to the class I heavy chain of interest.

4.1 cDNAs

Many murine, and human class I cDNAs have been isolated to date, and their sequences are readily available through the Genbank Database maintained at the National Library of Medicine at the NIH. These cDNAs should be obtainable from the original reporting laboratories, or can be easily cloned by PCR using flanking primers based on their published sequences, using reverse transcribed RNA derived from an appropriate animal tissue, or cell line as template. If EMCV-based vectors such as pCITE are used (see above), the class I 5′UT must be replaced with that of the EMCV (which is contained in the plasmid), and as such it will be convenient to create an *Nco*I site (*Nco*I recognizes the sequence CCATGG) at the initiation ATG of the cDNA to facilitate cloning in-frame at the ATG of the plasmid.

4.2 Antibodies

The generation of numerous haplotype-specific antibodies to murine class I molecules has greatly facilitated the study of class I MHC biology. Within the sets of antibodies to specific class I alleles, there are antibodies to defined domains, which have allowed the discrimination of unfolded, partially folded, and natively folded proteins. This is a unique feature of the murine system. Fewer mAbs exist that are so specific or well characterized for the study of human class I proteins, although more and more are being developed and reported. Many of these mAb hybridomas are available through the American Type Culture Collection (Rockville, Md). A partial list of some of the more common antibodies to murine and human class I epitopes are listed below in *Table 1*.

4.3 Peptides

Two major contributions to the identification of peptide binding motifs for class I molecules have been the solution of the crystal structure for more and more class I molecules, and the ability to elute and sequence bound peptides directly from immunoaffinity purified class I molecules isolated from cells. This latter advancement, which traditionally involved the acid elution of peptides from the class I molecules followed by HPLC separation and microsequencing has now been further adapted to more efficiently identify these peptides. Briefly, HPLC fractions from eluted class I molecules are directly analysed by tandem mass spectrometry to determine the peptide sequence, which requires significantly less protein, and takes less time to complete (39). Once these peptides have been identified, synthetic analogues can be synthesized and used for studies of assembly. This has led to the identification of binding motifs for many class I MHC molecules (40–51).

Table 1. Common antibodies to murine and human class I epitopes

Antibody	**Specificify**	**Cross-reaction**	**Class**	**Reference**
11–4.1	K^k	p, q, r	IgG2a	24
3–83	K^k, D^k	$K^{b,s,p,q,r}$	IgG2a, k	25
12–2–2	K^k, D^k	$K^{q,p,r}$	IgM, k	25
15–5–5	D^k	$K^{d,f}$	IgG2a, k	25
31–3–4	K^d	None	IgM	26
20–8–4	K^d, D^d	K^b, $D^{b,r,s}$		27
34–1–2	K^d, D^d	$K^{b,s,p,q,r}$	IgG2a, k	27
27–11–13	D^d	$D^{b,q}$	IgG2a, k	27
34–2–12	D^d (α3)	None	IgG2a, k	26
34–4–20	D^d	K^b, $K^{q,r}$	IgG2a, k	26
34–5–8	D^d ($\alpha1\alpha2$)	None	IgG2a, k	26
34–4–21	D^d	None	IgM	26
23–5–21	D^d	$D^{b,q,p}$	IgM, k	27
15–1–5	D^d	K^k, D^k	IgG2b, k	25
28–11–5	D^d (α2)	$D^{b,q,p}$	IgM, k	27
28–8–6	D^d	K^b, D^b	IgG2a, k	27
30–5–7	L^d (α2)	D^q, L^q	IgG2a, k	28
28–14–8	L^d (α3)	D^b, D^q, L^q	IgG2a, k	28
23–10–1	L^d	D^q, L^q	IgM, l	28
64–3–7	L^d (α2)			29
34–7–23	K^d, D^d	K^b,q	IgG2a, k	26
28–13–3	K^b	f	IgM, k	27
34–2–20	K^b	D^k, K^q	IgG2a, k	26
Y-3	K^b, K^k		IgG2b	30
B22.249	D^b, α1		IgG2a, k	31
7–2–14	D^d COOH term.		Rat IgM	32
W6/32	HLA A, B, C		IgG2a	33
MA2.1	HLA A.2, B–17			34
ME1	HLA B–7, B–27, Bw22 (α1)		IgG1	35
BB7.2	HLA A–2 (α1)		IgG2b	36
BBM1	Human β2-m		IgG2b	37
S-19	Murine $\beta 2pm^b$			38

5. Glycosylation of MHC class I

The extracelluar domains of class I MHC molecules contain carbohydrate residues. Two glycosylation sites, at positions 86 and 176 are highly conserved throughout mammalian species. Some class I MHC molecules, such as H2-L^d have a third glycosylation site in the α3 domain at position 256. Despite this conservation of these sites, and the ubiquitous nature of sugar moieties on the majority of membrane proteins, their precise biological significance has not been fully determined. Site-directed mutagenesis of class I MHC genes to alter the canonical glycosylation sequence, Asn–Xxx–Thr, or Asn–Xxx–Ser resulting in substitution of either Gln or Lys for Asn has revealed a significant reduction in surface expression, but no difference in serological determinants, or the ability to be recognized by alloreactive CTL (52). These data were interpreted to imply that the biological significance of glycosylation of class I molecules may lie in the efficiency of intracellular transport. Results of experiments in which glycosylation is inhibited either by use of sugar analogues, treatment with tunicamycin, or glycosidases are often difficult to interpret due to the potential effects of these agents on other cellular components. Investigations of assembly of mutagenized class I molecules using the *in vitro* translation and assembly should provide information on the role of core glycosylation in efficient association of the chains into mature complexes.

References

1. Krangel, M. S., Orr, H. T., and Strominger, J. L. (1979). *Cell*, **18**, 979.
2. Jackson, M. R., Cohen-Doyle, M. F., Peterson, P. A., and Williams, D. B. (1994). *Science*, **263**, 384.
3. Ahluwalia, N., Bergeron, J., Wada, I., Degen, E., and Williams, D. B. (1992). *J. Biol. Chem.*, **267**, 10914.
4. Christinck, E. R., Luscher, M. A., Barber, B. H., and Wiliams, D. B. (1991). *Nature*, **352**, 67.
5. Degen, E. and Williams, D. B. (1991). *J. Cell. Biol.*, **112**, 1099.
6. Dobberstein, B., Garoff, H., Warren, G., and Robinson, P. J. (1979). *Cell*, **17**, 759.
7. Kvist, S. and Hamann, U. (1990). *Nature*, **348**, 446.
8. Levy, F., Gabathuler, R., Larsson, R., and Kvist, S. (1991). *Cell*, **67**, 265.
9. Levy, F., Larsson, R., and Kvist, S. (1991). *J. Cell. Biol.*, **115**, 959.
10. Ribaudo, R. K. and Margulies, D. H. (1992). *J. Immunol.*, **149**, 2935.
11. Bijlmakers, M. J., Neefjes, J. J., Wojcik-Jacobs, E. H., and Ploegh, H. L. (1993). *Eur. J. Immunol.*, **23**, 1305.
12. Pelham, H. R. and Jackson, R. J. (1976). *Eur. J. Biochem.*, **67**, 247.
13. Shepherd, J. C., Schumacher, T. N., Ashton-Rickardt, P. G., Imaeda, S., Ploegh, H. L., Janeway, C. A. J., *et al.* (1993). *Cell*, **74**, 74577.
14. Bulleid, N. J. and Freedman, R. B. (1988). *Nature*, **335**, 649.
15. Freedman, R. B., Brockway, B. E., and Lambert, N. (1984). *Biochem. Soc. Trans.*, **12**, 929.
16. Hillson, D. A., Lambert, N., and Freedman, R. B. (1984). In *Methods in Enzymology*, **107**, 281–94. F. Wold and K. Moldave (eds.). Acad. Press Inc., N.Y.
17. Hwang, C., Sinskey, A. J., and Lodish, H. F. (1992). *Science*, **257**, 1496.
18. Scheele, G. and Jacoby, R. (1982). *J. Biol. Chem.*, **257**, 12277.
19. Kozak M. (1987). *Nucleic Acids Res.*, **15**, 8125.
20. Jackson, R. J. (1991). *Biochim. Biophys. Acta*, **1088**, 345.
21. Blobel, G. and Dobberstein, B. (1975). *J. Cell. Biol.*, **67**, 835.
22. Walter, P. and Blobel, G. (1981). *J. Cell. Biol.*, **91**, 557.
23. Walter, P. and Blobel, G. (1983). *Methods in Enzymology*, **96**, 84–93. S. Fleischer and B. Fleischer (eds.). Acad. Press Inc., N.Y.

24. Oi, V. T., Jones, P. P., Godling, J. W., Herzenberg, L. A., and Herzenberg, L. A. (1978). *Curr. Top. Microbiol. Immunol.*, **81**, 115.
25. Ozato, K., Mayer, N., and Sachs, D. H. (1980). *J. Immunol.*, **124**, 533.
26. Ozato, K., Mayer, N., M., and Sachs, D. H. (1982). *Transplantation*, **34**, 113.
27. Ozato, K. and Sachs, D. H. (1981). *J. Immunol.*, **126**, 317.
28. Ozato, K., Hansen, T. H., and Sachs, D. H. (1980). *J. Immunol.*, **125**, 2473.
29. Lie, W. R., Myers, N. B., Connolly, J. M., Gorka, J., Lee, D. R., and Hansen, T. H. (1991). *J. Exp. Med.*, **173**, 449.
30. Jones, B. and Janeway, C. A. J. (1981). *Nature*, **292**, 547.
31. Hammerling, G. J., Hammerling, U., and Flaherty, L. (1979). *J. Exp. Med.*, **150**, 108.
32. Southern, S. O., Swain, S. L., and Dutton, R. W. (1989). *J. Immunol.*, **142**, 336.
33. Kahn-Perles, B., Boyer, C., Arnold, B., Sanderson, A. R., Ferrier, P., and Lemonnier, F. A. (1987). *J. Immunol.*, **138**, 2190.
34. Ways, J. P., Rothbard, J. B., and Parham, P. (1986). *J. Immunol.*, **137**, 217.
35. Ellis, S. A., Taylor, C., and McMichael, A. (1982). *Hum. Immunol.*, **5**, 49.
36. Parham, P. and Brodsky, F. M. (1981). *Hum. Immunol.*, **3**, 277.
37. Brodsky, F. M., Bodmer, W. F., and Parham, P. (1979). *Eur. J. Immunol.*, **9**, 536.
38. Tada, N., Kimura, S., Hatzfeld, A., and Hammerling, U. (1980). *Immunogenetics*, **11**, 441.
39. Hunt, D. F., Henderson, R. A., Shabanowitz, J., Sakaguchi, K., Michel, H., Sevilir, N., *et al.* (1992). *Science*, **255**, 1261.
40. Corr, M., Boyd, L. F., Frankel, S. R., Kozlowski, S., Padlan, E. A., and Margulies, D. H., (1992). *J. Exp. Med.*, **176**, 1681.
41. Deres, K., Beck, W., Faath, S., Jung, G., and Rammensee, H. G. (1993). *Cell. Immunol.*, **151**, 158.
42. Falk, K., Rotzschke, O., Stevanovic, S., Jung, G., and Rammensee, H. G. (1991). *Nature*, **351**, 290.
43. Falk, K., Rotzschke, O., Grahovac, B., Schendel, D., Stevanovic, S., Gnau, V. *et al.* (1993). *Proc. Natl. Acad. Sci. USA*, **90**, 12005.
44. Falk, K., Rotzschke, O., Grahovac, B., Schendel, D., Stevanovic, S., Jung, G. *et al.* (1993). *Immunogenetics*, **38**, 161.
45. Falk, K., Rotzschke, O., Stevanovic, S., Jung, G., and Rammensee, H. G. (1994). *Immunogenetics*, **39**, 230.
46. Malcherek, G., Falk, K., Rotzschke, O., Rammensee, H. G., Stevanovic, S., Gnau, V. *et al.* (1993). *Int. Immunol.*, **5**, 1229.
47. Norda, M., Falk, K., Rotzschke, O., Stevanovic, S., Jung, G. and Rammensee, D. H. (1993). *J. Immunol.*, **14**, 144.
48. Rammensee, H. G., Falk, K., and Rotzschke, O. (1993). *Annu. Rev. Immunol.*, **11**, 213.
49. Rotzschke, O., Falk, K., Deres, K., Schild, H., Norda, M., Metzger, J., *et al.* (1990). *Nature*, **348**, 252.
50. Rotzschke, O., Falk, K., Stevanovic, S., Jung, G., and Rammensee, H. G. (1992). *Eur. J. Immunol.*, **22**, 2453.
51. Wallny, H. J., Deres, K., Faath, S., Jung, G., Van Pel, A., Boon, T. *et al.* (1992). *Int. Immunol.*, **4**, 1085.
52. Miyazaki, J., Appella, E., Zhao, H., Forman, J., and Ozato, K. (1986). *J. Exp. Med.*, **163**, 856.

4

Expression, purification, and characterization of recombinant soluble MHC class I molecules

ANDERS BRUNMARK and MICHAEL R. JACKSON

1. Introduction

Functional MHC class I molecules are made up of three components, transmembrane MHC class I heavy chain, soluble β_2-microglobulin, and a short peptide which is typically eight to ten amino acids long. This trimolecular complex is assembled within the endoplasmic reticulum and transported to the cell surface to which it is anchored via the transmembrane domain. Initial attempts to solve the structure of this complex utilized proteolysis to release a soluble class I fragment from the surface of human cells. Using this material the structure of the extracelluar domain of an MHC class I molecule containing a heterogeneous mixture of peptides was solved in 1987 (1). Since then many laboratories throughout the world have sought ways to express recombinant soluble MHC class I molecules, either devoid of peptide or containing a single peptide species, with the goal that this material would allow the determination of:

- three-dimensional structure of a class I containing a single peptide species
- binding constants of peptides for class I
- rules that govern peptide binding and thermostability of class I complexes
- understanding of T cell recognition of MHC class I

Production of recombinant soluble class I appeared straightforward. Expression plasmids containing cDNA encoding truncated soluble class I and β_2-microglobulin were easily generated and many excellent expression system in a variety of organisms were available. However, production was to prove much more difficult than anticipated in all the standard systems. Key to production was the recognition that empty class I molecules unlike their peptide-filled counterparts are thermally unstable, i.e. they disintegrate at 37 °C but are still reasonably stable at 24–26 °C (2).

1.1 Expression in mammalian cells

The problems with expression in mammalian cells are:

(a) Mammalian cells grow best at 37 °C, a temperature at which empty class I disintegrate.
(b) Cell lines used for high level expression of recombinant protein (e.g. CHO) load peptide on to at least some of the class I produced.
(c) Mammalian cells transport empty class I out of the endoplasmic reticulum very slowly (especially HLA).
(d) Empty class I readily pick up peptide and exchange β_2-microglobulin once secreted into the medium necessitating serum-free medium.

Despite these problems Fahnestock *et al.* (3) have expressed soluble K^d in CHO cells using a glutamine synthetase amplifiable expression system. Reasonably high levels of soluble class I were secreted especially if human β_2-microglobulin was co-expressed (the higher affinity of human β_2-microglobulin for murine class I imparts thermostability to the empty complexes). However, approximately 30% of the molecules secreted in this system contained high affinity peptides, and as empty (or low affinity peptide) and peptide-containing molecules are difficult to separate this severely limits the use of the system. Nevertheless the protein produced can be denatured and refolded if high affinity peptide or human β_2-microglobulin is present.

1.2 Expression in bacteria

Class I heavy chains and β_2-microglobulin expressed in bacteria are typically denatured and in most cell types sequestered into inclusion bodies. Successful refolding of this bacterially expressed material by dilution out of denaturing conditions was found to require the presence of a high affinity peptide (4, 5). Although this limits the use of bacterial expression— as empty molecules *per se* are not generated, bacterial expression is undoubtedly useful for the production of large amounts of crystallography grade class I molecules containing a single peptide species (5, 6). The advantages of this method are that it is relatively cheap and rapid to set-up and can produce large amounts of material. However it is worth noting that some class I molecules, e.g. HLA-B27 are much more difficult to refold than others, e.g. HLA-A2.1.

1.3 Expression in insect cells

From the outset the baculoviral system would seem ideally suited to production of recombinant empty soluble class I. Not only is this system reknown for its high levels of production but the insect host cells derived from *Spodoptera frugiperda* grow well in defined serum-free medium at 26 °C, a temperature at which the empty molecules are relatively stable. Despite the effort of many groups, the yield of correctly folded class I from this system has been found to be very poor since only a small fraction (< 5%) of the class I heavy chains and β_2-microglobulin were found to assemble into complexes (7). The reasons for this poor assembly are not known. Nevertheless insect cells are an attractive host cell as:

(a) They grow at room temperature.
(b) They efficiently assemble, glycosylate, and secrete molecules.
(c) Expressed class I are unlikely to be loaded with peptide as insects do not have MHC class I and are therefore unlikely to have the machinery to generate/load peptide.

In order to take advantage of these characteristics we have used *Drosophila*-based expression system (8, 9) in which cell lines with stably integrated copies of class I heavy chain and β_2-microglobulin genes are established. The major advantages of this system are that:

(a) User friendly expression vectors with tightly controlled inducible promoters are available.
(b) Stably transfected cell lines are easily generated and maintained.
(c) Empty soluble class I molecule of all haplotypes tested (20 to date) are secreted from the cells at levels of 0.5–2 mg/litre.

The major disadvantage of the system is that expression levels are only moderate and thus production costs are quite high.

2. Expression of recombinant proteins in *Drosophila* cells

In the following section we will describe this *Drosophila* expression system, the expression plasmid used, how to transfect the *Drosophila* cells, and how to generate stably transfected cell lines. Although the methodology described focuses on the production of soluble class I, the same strategy is readily applicable to the production and purification of other proteins ordinarily expressed at the cell surface as well as truncated, soluble proteins. Indeed we have found this system amazingly versatile for the production of many other soluble recombinant proteins, e.g. MHC class II, invariant chain, T cell receptor, CD4, CD8, B7, I-CAM. In many cases the exact same constructs were not transported or expressed in mammalian cells. The reason why such molecules are transported in *Drosophila* cells but not mammalian cells is not clear, but recent data suggests that in *Drosophila* cells the recombinant molecules may evade the normal quality control exerted by the molecular chaperone calnexin (10).

2.1 *Drosophila* expression plasmids

Although a variety of expression vectors are available for transfection of *Drosophila* cells, one with an inducible promoter is advantageous when

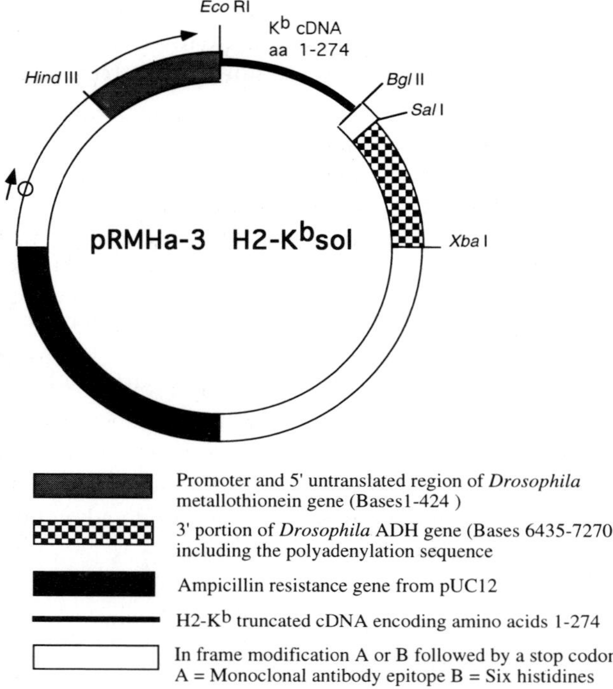

Figure 1. Schematic map of pRMHa-3 containing a cDNA encoding soluble H2-K^b followed by histidine $(His)_6$ tag sequence.

generating stably transfected cell lines. The metallothionine promoter is preferable to the heat shock promoter and good results have been obtained with the expression plasmid pRMHa-3 (11), (*Figure 1*). This plasmid has a polylinker sequence containing multiple restriction enzyme sites located immediately downstream of a metallothionine promoter and upstream of a polyadenylation sequence. The rest of the plasmid is essentially the workhorse plasmid pUC12. We have observed no site preference in this polylinker for maximal expression of the recombinant protein.

In order to ensure expression of the cDNA all 5′ non-coding DNA sequence should be removed and the cDNA should be engineered to begin with a Kozak consensus sequence (CCACC**ATG**) where the ATG encodes the initiator methionine (12). High levels of expression have been obtained with many cDNAs in which the 5′ non-coding sequence was maintained (up to 100 bp) however for other cDNA we have found it essential to remove all 5′ non-coding sequence to obtain any expression. Truncating transmembrane proteins to obtain a soluble recombinant molecule is achieved by simply introducing a stop codon immediately after the last residue before the transmembrane domain. In order to aid in purification and identification of the recombinant protein the stop codon can be preceded by an epitope tag consisting of a recognition sequence for a monoclonal antibody (13, 14) and/or a series of six histidines for nickel agarose purification (15). Examples of epitope tags are a peptide sequence from influenza haemagglutnin—YPYDVPDYA (13) and from the proto-oncogene *c-myc*—EQKLISEEDL (14). Monoclonal antibodies that recognize these epitopes (12CA5 and 9E10) are available. For reference *Figure 2* shows the complete amino acid sequence of the soluble $(His)_6$ tagged H2-K^b encoded by the plasmid in *Figure 1*. β_2-microglobulin cDNA is similarly cloned into pRMHa-3 without epitope tagging. Note that if this system is being used to express a heterodimer, engineering different tags on to each chain can greatly aid in purification of heterodimers. The $(His)_6$ tag has proven to be especially useful, purification of soluble empty class I using nickel NTA–agarose (Ni–agarose) chromatography results in much higher yields of correctly folded protein compared to antibody affinity purification. The latter requires elution of the class I from the column under rather harsh conditions, leading to a loss of greater than 80% of the empty class I (9).

2.2 Recombinant DNA strategies for generating expression plasmids

Space does not allow us to describe in detail the recombinant DNA methods used to carry out the construction of the plasmid described above. However all the techniques required are standard and described in many excellent text books on this subject (16). Below we have outlined the general strategy to generate the recombinant plasmids required.

(a) Starting with cDNA encoding a full-length class I molecule the polymerase chain reaction (PCR) is the most convenient way of tailoring the ends of the molecule. Due to PCR errors the use of vent or other proof-reading polymerase is highly recommended. If PCR errors are found within the main bulk of the molecule it is often easiest to remove the mutated region as a restriction fragment and replace it with the same fragment from the original cDNA. The following PCR primers:

    5′ primer TTTGAATTCACCATGGTACCGTGCACGCTGCTCCT
    3′ primer TTTAGATCTCCATCTCAGGGTGAGGGGCTC

    were used to generate the PCR fragment for K^bsol $(His)_6$. The PCR products were purified using the GeneClean kit (Bio 101), cleaved with *Eco*RI and *Bgl*II, and ligated into these sites in the pRMHa-3 $(His)_6$ tag vector—see below.

(b) Epitope tags are easily generated by directionally cloning a pair of complimentary oligoucleotides with appropriate sticky-end overhangs into the pRMHa-3 polylinker. The following pair of oligonucleotides:

    5′ AATTCAGATCTCACCATCACCATCACCATTGAG 3′
    5′ TCGACTCAATGGTGATGGTGATGGTGAGATCTG 3′

```
EcoRI
GAATTCACC ATG GTA CCG TGC ACG CTG CTC CTG CTG TTG GCG GCC    45
           M   V   P   C   T   L   L   L   L   L   A   A    -10

GCC CTG GCT CCG ACT CAG ACC CGC GCG GGC CCA CAC TCG CTG AGG  90
A   L   A   P   T   Q   T   R   A   G   P   H   S   L   R    6

TAT TTC GTC ACC GCC GTG TCC CGG CCC GGC CTC GGG GAG CCC CGG  135
Y   F   V   T   A   V   S   R   P   G   L   G   E   P   R    21

TAC ATG GAA GTC GGC TAC GTG GAC GAC ACG GAG TTC GTG CGC TTC  180
Y   M   E   V   G   Y   V   D   D   T   E   F   V   R   F    36

GAC AGC GAC GCG GAG AAT CCG AGA TAT GAG CCG CGG GCG CGG TGG  225
D   S   D   A   E   N   P   R   Y   E   P   R   A   R   W    51

ATG GAG CAG GAG GGG CCC GAG TAT TGG GAG CGG GAG ACA CAG AAA  270
M   E   Q   E   G   P   E   Y   W   E   R   E   T   Q   K    66

GCC AAG GGC AAT GAG CAG AGT TTC CGA GTG GAC CTG AGG ACC CTG  315
A   K   G   N   E   Q   S   F   R   V   D   L   R   T   L    81

CTC GGC TAC TAC AAC CAG AGC AAG GGC GGC TCT CAC ACT ATT CAG  360
L   G   Y   Y   N   Q   S   K   G   G   S   H   T   I   Q    96

GTG ATC TCT GGC TGT GAA GTG GGG TCC GAC GGG CGA CTC CTC CGC  405
V   I   S   G   C   E   V   G   S   D   G   R   L   L   R    111

GGG TAC CAG CAG TAC GCC TAC GAC GGC TGC GAT TAC ATC GCC CTG  450
G   Y   Q   Q   Y   A   Y   D   G   C   D   Y   I   A   L    126

AAC GAA GAC CTG AAA ACG TGG ACG GCG GCG GAC ATG GCG GCG CTG  495
N   E   D   L   K   T   W   T   A   A   D   M   A   A   L    141

ATC ACC AAA CAC AAG TGG GAG CAG GCT GGT GAA GCA GAG AGA CTC  540
I   T   K   H   K   W   E   Q   A   G   E   A   E   R   L    156

AGG GCC TAC CTG GAG GGC ACG TGC GTG GAG TGG CTC CGC AGA TAC  585
R   A   Y   L   E   G   T   C   V   E   W   L   R   R   Y    171

CTG AAG AAC GGG AAC GCG ACG CTG CTG CGC ACA GAT TCC CCA AAG  630
L   K   N   G   N   A   T   L   L   R   T   D   S   P   K    186

GCC CAT GTG ACC CAT CAC AGC AGA CCT GAA GAT AAA GTC ACC CTG  675
A   H   V   T   H   H   S   R   P   E   D   K   V   T   L    201

AGG TGC TGG GCC CTG GGC TTC TAC CCT GCT GAC ATC ACC CTG ACC  720
R   C   W   A   L   G   F   Y   P   A   D   I   T   L   T    216

TGG CAG TTG AAT GGG GAG GAG TTG ATC CAG GAC ATG GAG CTT GTG  765
W   Q   L   N   G   E   E   L   I   Q   D   M   E   L   V    231

GAG ACC AGG CCT GCA GGG GAT GGA ACC TTC CAG AAG TGG GCA TCT  810
E   T   R   P   A   G   D   G   T   F   Q   K   W   A   S    246

GTG GTG GTG CCT CTT GGG AAG GAG CAG TAT TAC ACA TGC CAT GTG  855
V   V   V   P   L   G   K   E   Q   Y   Y   T   C   H   V    261
                                                BglII
TAC CAT CAG GGG CTG CCT GAG CCC CTC ACC CTG AGA TGG AGA TCT  900
Y   H   Q   G   L   P   E   P   L   T   L   R   W   R   S    276
                            SalI
CAC CAT CAC CAT CAC CAT TGA GTCGAC                           927
H   H   H   H   H   H   *                                    282
```

Figure 2. Full sequence and translation of cDNA encoding recombinant H2-K^bsol $(His)_6$. This cDNA is then cloned between *Eco* RI and *Sal*I sites in the polylinker of pRMHA-3. Note the inclusion of a Kozak consensus sequence around the initiator methionine, a restriction site (*Bgl*II) at the end of the alpha 3 domain, and the $(His)_6$ tag preceding the stop codon.

were cloned between the *Eco*RI and *Sal*I in the pRMHa-3 polylinker generating the $(His)_6$ tag for K^bsol $(His)_6$ shown in *Figure 1*.

(c) Even if you do not want to epitope tag your soluble class I it is highly advisable to include a unique restriction enzyme site immediately before the stop codon. In this way a second round of PCR is not required should you change your mind!

Recombinant plasmids should be completely sequenced using either internal primers and/or 5′ and 3′ pRMHa-3 primers. The following 5′ and 3′ pRMHa-3 primers work well for both manual and semi-automated dideoxy sequencing.

5′pRMHa-3 sequencing primer GAGCAGCAAAATCAAGT
3′pRMHa-3 sequencing primer GAAGAATGTGAGTGTGC

Once the sequence is verified large scale preparation of DNA using the standard alkaline lysis method followed by CsCl gradient centrifugation (one gradient is sufficient) or column chromatography (e.g. Quiagen magic maxi-preps) is required to obtain high quality transfection grade DNA. The pRMHa-3 vector is high-yielding; a 100 ml maxi-prep typically provides 2–5 mg of DNA. Purified DNA should be finally precipitated from 70% EtOH and resuspended in sterile water. DNA concentration should be accurately determined spectrophotometrically.

2.3 *Drosophila* Schneider cells—general maintenance

2.3.1 Introduction

Drosophila Schneider cells grow at room temperature (22–25 °C) in closed capped flasks in Schneider medium (Gibco BRL) supplemented with 10% fetal calf serum (heat inactivated for 1 h at 55 °C) hereafter referred to as complete Schneiders medium. The surface to volume ratio in the flask should be no more than 2 ml/10 cm^2, i.e. no more than 15 ml of media in a 75 cm^2 flask. Cells should be maintained at a cell density of 0.5–10 $\times$ 10^6 although they will grow to much higher cell densities and remain viable.
You need:

- tissue culture hood
- sterile 1 ml and 10 ml pipettes
- sterile 75 cm^2 tissue culture flasks
- Schneiders complete medium
- fetal calf serum plus 10% dimethyl sulfoxide
- clinical bench-top centrifuge

2.3.2 Passaging cells

Gently tap side of flask containing a dense culture of cells (1 $\times$ 10^7 cells/ml) to displace adherent cells, transfer 1 ml of cells into a fresh flask containing 14 ml of Schneiders complete media. Keep the old flask as a back-up. Viable cells can be recovered from such back-ups even after two to three weeks. Cell cultures passaged in this way have been maintained in continuous culture for more than a year. Stably transfected cell lines should be maintained in the presence of G418 at a concentration of 500 μg/ml (active).

2.3.3 Freezing cells

Remove 10 ml of cells in exponential growth phase (4–8 $\times 10^6$ cells/ml) from the flask into a 15 ml sterile centrifuge tube. Collect cells by centrifugation at 1000 *g* for 3 min, discard supernatant, resuspend cells in 2 ml of cold cell freezing media (90% FCS, 10% DMSO), aliquot into liquid N_2 freezer tube. Place cells at −70 °C for one to three days, then transfer to liquid N_2 storage. Cells maintained at −70 °C have remained viable for more than one year. Cells in liquid nitrogen have been recovered after three years storage.

2.4 Transfection of Schneider cells

Schneider cells are very easily transfected using the $CaPO_4$ technique. Typical efficiencies for transient transfections are 20–40%. $CaPO_4$/DNA precipitates do not appear to be at all toxic to the Schneider cells and precipitates can be left on cells for 48 h or more. Multiple copies of DNA are taken up and integrated by each transfected cell and no problems are encountered for co-transfections. Indeed, six different plasmids have been successfully co-transfected and stably integrated at high efficiency into the same cell using this technique. The protocol for transient transfections is essentially the same as for generating stables up to day four when $CuSO_4$ (final conc. 1 mM) rather than G418 is added to induce expression. Note that for bulk protein production we have found that the use of cloned cells expressing high levels of recombinant protein can often increase yield by one order of magnitude.

Protocol 1. Transfection of Schneider cells

Equipment and reagents

- Plasmid DNA: purified supercoiled DNA in H_2O
- Schneiders complete medium
- Sterile Ca^{2+} solution: 500 mM $CaCl_2$, 100 mM Hepes (pH 6.7 at 37 °C)
- Geneticin sulfate (Gibco): 100 mg/ml G418 (200 × stock)
- Sterile phosphate solution: 280 mM NaCl, 0.75 mM Na_2HPO_4, 0.75 mM NaH_2PO_4, 50 mM Hepes (pH 6.7 at 37 °C)
- 100 cm^2 tissue culture Petri dishes
- 162 cm^2 cell culture flasks
- Clinical low speed centrifuge
- Tissue culture hood

Method

1. Day 0. Dilute a culture of healthy Schneider cells (at density 3–8 $\times 10^6$/ml) to 1 $\times 10^6$/ml. Pipette 10 ml of diluted cells into a 100 mm Petri dish. Healthy cells will attach to the plastic within 1 h.

2. Day 1. Pipette 25 μg of supercoiled plasmid DNA, e.g. 12 μg H2-K^bsol $(His)_6$ pRMHa-3, and 12 μg of β_2-microglobulin pRMHa-3, and 1 μg of the neomycin resistance plasmid (phshsneo) (17) into a 4 ml sterile capped tube. Make the volume up to 240 μl with sterile H_2O, vortex.
3. Place tube in 37 °C heater block along with tubes containing 0.2 μm filter sterilized Ca^{2+} and phosphate solutions. The following reactions are then carried out in the heater block and added to the Schneider cells in a tissue culture hood.
4. Add 240 μl of Ca^{2+} solution to DNA and vortex for 10 sec. Leave in heater block for 10 min.
5. Add 480 μl phosphate solution, vortex for 10 sec. Leave for an additional 15 min in the heater block.
6. Add solution dropwise on to cells. Upon addition the media should go slightly cloudy. Wrap Petri dishes in Cling Film to prevent evaporation and leave for two days.
7. Day 3. Remove cells from Petri dish by pipetting up and down and transfer to a 15 ml tube. Pellet cells by centrifugation at 400 *g* for 3 min. Aspirate supernatant and transfer cells into 20 ml of Schneiders media in a 162 cm^2 flask. Leave cells for one day.
8. Day 4. To select for stably transfected cells G418 should be added directly into the flask to a final concentration of 500 μg/ml. For transient expression add $CuSO_4$ at this point to a final concentration of 1 mM and analyse cells for expression of the recombinant protein on day 5 or 6 (see below). Note that the heat shock promoter in the phshsneo plasmid is leaky and heat shock is not required to switch it on.
9. Days 10–40. Every five to seven days dilute cells 1:3 to 1:10 by transferring the required volume of cells into a fresh 162 cm^2 flasks containing 15 ml of Schneider complete medium containing 500 μg/ml G418. Keep the old culture flasks as back-ups. Schneider cells are extremely hardy and can be removed from these back-ups many weeks later! During the selection procedure dividing cells usually form clusters (good sign) whereas big cells containing large vacuoles are dying. Do not split unless cells are clearly dividing and getting more dense. After multiple rounds of dilution (three to four weeks) the culture will consist of a fairly homogeneous culture of G418-resistant stably transfected cells which upon dilution readily adhere to fresh plastic surfaces.

Note: always check to determine that your recombinant protein is expressed under transient expression (see below), there is no point waiting for the stable cell lines to find out your construct does not work. If after six weeks of selection a stable population of G418-resistant cells is not obtained you may have over-diluted the cells. The effective concentration of G418 is somewhat dependent upon the cell density/number of G418-resistant cells.

3. Analysis of transfected Schneider cells

Transient and stable populations of cells can be checked for expression of the recombinant protein by a variety of techniques. If the recombinant molecule is soluble and secreted the most direct method is to carry out a mini-scale purification and check directly that the protein is expressed and secreted. For analysis of transients expression levels are typically low such that metabolic radiolabelling of cells followed by immunoprecipitation from the supernatant and cells or a sensitive Western blot analysis of the cells and supernatant is required (8, 10). However expression levels from stably transfected cells are typically sufficiently high that the affinity purified protein can be visualized on a Coomassie stained SDS–PAGE gel. If the protein is expressed on the cell surface they can be stained using antibodies and analysed using a cytofluorometer. Alternatively transfection efficiency can be monitored by immunofluorescence microscopy. Obviously if the recombinant protein is efficiently secreted this is not a very quantitative method, nevertheless our experience is that some cells will still be observed as positive.

3.1 Pulse chase metabolic labelling of Schneider cells

This is the most reliable method in our hands to both detect expression of recombinant proteins and to check that they are efficiently transported out of the cells. This technique also allows one to determine if the class I molecules expressed are thermally unstable.

Protocol 2. Metabolic labelling of Schneider cells

Equipment and reagents

- Radioactive workspace
- [^{35}S]methionine (> 800 Ci/mmol) (Amersham)
- Schneiders complete medium
- Graces insect medium without methionine (Gibco BRL)
- PBS, PBS with 1% Triton X-100, 10 × PBS with 10% Triton X-100
- Protein A–Sepharose antibody (Pharmacia)
- Eppendorf microcentrifuge and bench-top clinical low speed centrifuge

Method

Note: before you can carry out this protocol you must receive appropriate training in the use of radioisotopes.

1. 12–24 h prior to labelling add 1 mM $CuSO_4$ to cells.
2. Collect $CuSO_4$ treated cells by centrifugation at 400 *g* for 3 min, resuspend in methionine-free Graces medium (5×10^7 cells/ml), and

incubate at room temperature for 10 min. Add [^{35}S]methionine to 0.5 mCi/ml and label for 15 min at room temperature. Add 9 vol. of Schneiders complete medium and remove (1 ml) aliquots of cells (5×10^6) at various chase times.

3. For each aliquot immediately separate cells from medium by a 10 sec spin in an Eppendorf microcentrifuge. Add 1 ml of PBS/1% Triton X-100 (4 °C) to cell pellet and vortex. Add 0.1 vol. of 10 × PBS/10% Triton X-100 to supernatant. Note that as empty class I is thermosensitive it is important to keep lysates cold throughout the immunoprecipitation procedure.
4. To pre-clear the lysates, centrifuge tubes at 12 000 *g* for 10 min, transfer supernatant to 1.5 ml Eppendorf containing 100 μl of a 10% slurry of protein A–Sepharose/PBS. Rotate tubes end over end for 1 h, spin 10 min at 12 000 *g* in a microcentrifuge, and transfer supernatant to a fresh tube.
5. In order to demonstrate that the class I molecules expressed are empty, lysates can be thermally challenged at this point by simply placing an aliquot at 37 °C for 1 h. As a control, add peptide (25 μM) known to bind to the class I to one aliquot prior to the thermal challenge.
6. Add antibody (1 μl of antisera or ascites) and 100 μl of a 10% slurry of protein A–Sepharose/PBS, rotate and spin tubes as above, wash Sepharose beads four times with 1 ml of PBS/0.1% Triton X-100.
7. After the final wash add 50 μl of SDS–PAGE sample buffer to pelleted beads, heat to 90 °C for 3 min, and analyse by SDS–PAGE (12.5% gels are suitable for analysis of soluble class I) and autoradiograph.

3.2 Immunofluorescence microscopy

This is the most rapid method to determine whether your recombinant protein is expressed and whether your transfection worked. It provides a good indication of the efficiency of transfection especially if a cell surface molecule is being analysed.

Protocol 3. Immunofluorescence microscopy

Equipment and reagents

- Fluorescence microscope
- Microscope slides and coverslips
- Fine forceps
- 6-well tissue culture plates
- Cell-Tak (Collaborative research)
- 0.1 M Na bicarbonate
- INSECT XPRESS medium (Biowhittaker)
- Quenching solution: 50 mM ammonium chloride in PBS
- Fixing solution: 3.7% formaldehyde in PBS
- Blocking buffer: 1% bovine serum albumin in PBS
- Permeabilization solution: 0.1% Triton X-100 in PBS
- Fluorescein labelled secondary antibody: FITC labelled affinity purified goat anti-mouse IgG
- Mounting media: Moviol (CalBiochem)

Protocol 3. *Continued*

Method

1. Place coverslips (20 × 20 mm) in 6-well plate, dilute Cell-Tak 1:20 with 0.1 M Na bicarbonate, and immediately pipette a 50 μl drop on to the centre of a coverslip, place a second coverslip on top.
2. Leave Cell-Tak sandwich for 15 min then add 3 ml of PBS and prise apart coverslip sandwich with fine-tipped forceps. Place coverslips coated side up in 6-well plate.
3. Collect $CuSO_4$ treated cells (1 × 10^6/coverslip) by centrifugation, re-suspend in 3 ml of XPRESS medium, add on to Cell-Tak treated coverslips. Cells should adhere within 20 min. Carefully aspirate XPRESS medium and add 3 ml of 3.7% formaldehyde solution. After 20 min aspirate media and carefully wash cells three times with 3 ml quenching solution. Permeabilize cells for 3 min with 3 ml 0.2% Triton X-100/PBS, then add 3 ml of blocking solution.
4. Invert coverslips on to a 50 μl drop of primary antibody (typically 1:500 dilution of ascites, 1:200 dilution antisera in blocking buffer) pipetted on to a piece of Parafilm. Incubate for 1 h.
5. Return coverslip to 6-well plate, wash three times with 3 ml blocking buffer, and invert on to a 50 μl drop of diluted FITC labelled secondary antibody (1:200). Incubate for 30 min, return coverslip to 6-well dish for three 3 ml washes with PBS. Finally rinse coverslip with H_2O and invert on to a drop of mounting medium. View under fluorescent microscope. A good transient transfection may be as high as 50% positive (e.g. for a cell surface molecule). For soluble constructs sufficiently high levels of accumulated intracellular molecules are typically observed in 5–20% of cells.

3.3 Mini-scale protein production (10–50 ml culture)

Mini-scale production is most suitable to test the potential of stable cell lines to secrete the recombinant class I, and in particular to test that the class I secreted binds nickel–agarose (Ni–agarose) or a particular antibody. Add $CuSO_4$ to 1 mM three days prior to harvest. Pellet cells (approx. 1 × 10^8 cells) from culture and transfer supernatant to a 15 ml tube. Add 100 μl of Ni–agarose (50% slurry in PBS) or protein A–Sepharose/Ab (10% slurry protein A/PBS pre-incubated with Ab) (3 μl), rotate end over end for 30 min, pellet beads 5 min 3000 *g*. Transfer beads to Eppendorf, wash twice with 1 ml PBS, and twice with 1 ml 20 mM imidazole/PBS pH 7.5. Analyse bound material by SDS–PAGE. If cells are expressing well a strong band of class I heavy chain and β_2-microglobulin should be visible by Coomassie staining. The identity of the protein isolated should be confirmed by Western blotting.

4. Large scale production of recombinant class I

Large scale production (2–15 litres) is most easily accomplished using roller bottles, however is also possible to carry out an intermediate level of production using multilayer tissue culture flasks. For production of homogeneous recombinant empty soluble class I it is obviously important to minimize contamination of the secreted molecules with short peptides from the media or by exchange of the β_2-microglobulin (β_2-m) with bovine β_2-m from serum. Use of serum-free media essentially removes the problem of heterologous β_2-m. However the problem of peptides in the culture media is more difficult to circumvent. All commercially available serum-free insect media contain a heterogeneous mixture of peptides derived primarily from yeastolate. These peptides are apparently essential for insect cells to grow. Clearly some of these peptides as well as peptides secreted by the *Drosophila* cells (e.g. degradation products) will be capable of binding class I. Isoelectric focusing and temperature challenge experiments on the purified class I obtained from this *Drosophila* expression system (9) showed that either the number of such occupied molecules is very low or that the affinity of the bound peptides was sufficiently low as to impart little thermostability to the complexes. Nevertheless, it is worth remembering that the serum-free medium is likely to contain a significant pool of these low affinity peptides. If the goal is to produce soluble class I containing a single peptide species then the problem of the empty class I picking up peptides from the media can be minimized by simply adding the desired peptide to the culture media (to a final concentration of 5 μM) at the time the cells are induced with $CuSO_4$. In general, addition of peptides to the medium significantly increases (two- to five-fold depending on the haplotype) the yield of class I obtained. We have noted that the different haplotypes of empty class I show differential thermosensitivity, the ways to improve the yields of the most thermosensitive molecules are:

(a) To reduce the temperature at which the insect cells are growing to 22 °C during the induction phase.

(b) To co-express the heavy chain with a heterologous β_2-m with higher affinity.

(c) To include a peptide that is known to bind to the class I in the media.

Protocol 4. Production of recombinant class I

Equipment and reagents

- Serum-free insect medium, e.g. INSECT XPRESS (Biowhittaker) or Excell 400 (JRH Bioscience)—warm to room temperature
- Roller bottle apparatus at 22–27 °C
- 850 cm^2 roller bottles
- 1 M $CuSO_4$, sterilized by filtration
- Tangential flow concentrator, e.g. Filtron concentrator with 30 kDa cut-off membrane

Protocol 4. *Continued*

Method

1. Dilute stably transfected population of *Drosophila* cells to 1×10^6/ml in Schneiders complete medium without G418, put 25 ml of diluted cells in a 162 cm^2 flask and leave for five days.
2. Transfer cells from above to a 850 cm^2 roller bottle. Add 75 ml XPRESS medium and put on the roller bottle apparatus, set at 1 r.p.m. After one to two days add 100 ml of XPRESS medium. It is important at this stage to check that the cells are adhering to the wall of the roller bottle. If this is not the case transfer cells to a new roller bottle.
3. Cells should now be growing exponentially and may be split 1:2 every one to three days. Maximum volume in the roller bottle is 250 ml. Maximum achievable cell density in XPRESS is around 10^7 cells/ml but cells are ready to be split when they have reached $4–8 \times 10^6$ cells/ml. Never dilute cells to $< 1 \times 10^6$/ml.
4. When the desired volume and density ($4–8 \times 10^6$ cells/ml) of cells have been obtained induce expression of the recombinant class I by adding $CuSO_4$ to 1 mM final concentration.
5. Incubate for an additional three days.
6. Spin down the cells at 400 *g* for 10 min.
7. Concentrate supernatant 20-fold using a tangential flow concentrator.
8. Cu^{2+} should be removed from the concentrate as it may interfere with protein binding to the Ni–agarose column. This can be achieved by overnight dialysis at 4°C against 20 vol. of PBS/0.02% sodium azide with two changes. Alternatively Cu^{2+} may be removed by filtration using the tangential flow apparatus by simply diluting the concentrate with PBS and re-concentrating several times. Add sodium azide to a final concentration of 0.02%.

5. Purification of soluble class I molecules

In the next section we describe the most straightforward method we have found to purify soluble class I molecules to homogeneity. Using only two purification steps, Ni–agarose affinity chromatography and anion exchange chromatography, preparations of class I are essentially homogeneous and free of contaminants. Material purified by these methods has been used for the determination of the structure of several class I–peptide complexes by X-ray crystallography. For reference *Figure 3* shows a Coomassie blue stained SDS–PAGE gel of fractions taken from various stages of the purification.

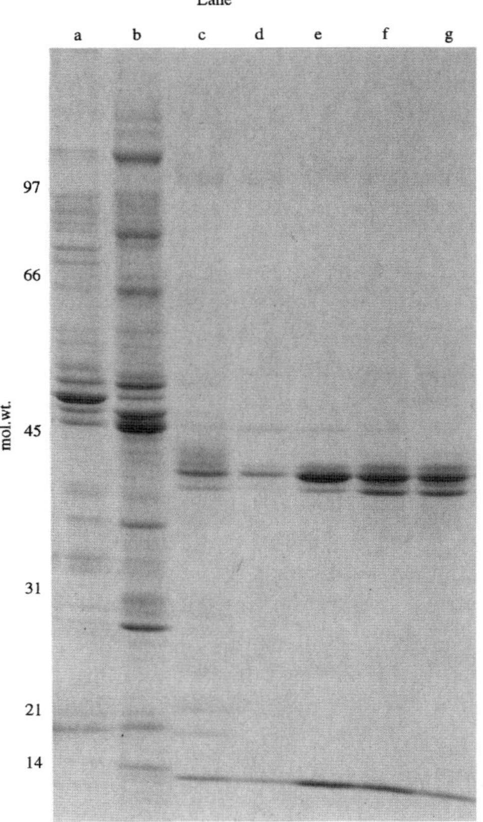

Figure 3. Purification of soluble D^bsol $(His)_6$. Coomassie stained 5–15% gradient SDS–PAGE gel of fractions at different stages of D^bsol $(His)_6$ purification. Lane a: concentrated supernatant (acetone precipitated 1/1000 of total loaded). Lane b: 20 mM imidazole wash from Ni–agarose column (*Figure 4*) (1/2000 loaded). Lane c: pooled peak fractions of material eluted from Ni–agarose column by 150 mM imidazole (*Figure 4*) (1/350 loaded). Lanes d, e, f, and g are fractions 1, 2, 3, and 4 from the anion exchange column purification see *Figure 5* (1/25 of each fraction loaded).

Obviously every protein has different properties requiring the modifications of the purification methods we describe. Nevertheless we have found that different class I molecules, both H2 and HLA as well as non-classical class I molecules, have remarkably similar biochemical properties. All empty class I molecules are relatively fragile and for this reason the mild elution conditions required for elution of class I from the Ni–agarose makes this a very much better purification step than antibody affinity purification.

Protocol 5. Nickel–agarose affinity column purification of class I molecules

Equipment and reagents

- Nickel NTA agarose beads (Quiagen)
- Low pressure chromatography column (Pharmacia)
- Peristaltic pump (Pharmacia P1 pump)
- UV detector reading at 280 nm (Pharmacia)
- PBS
- 20 mM imidazole in PBS, pH 7.4
- 150 mM imidazole in PBS, pH 7.4

Method

The following operations are carried out at 4 °C with cold solutions.

1. Centrifuge the concentrate at 10 000 *g* for 20 min to remove any aggregates/debris.
2. Load the concentrate on to a Ni–agarose column at a flow rate of 1–2 ml/min. A 3 ml bed volume of Ni–agarose is sufficient for a 5 litre preparation.
3. Wash the column with PBS until baseline absorbance is reached.
4. Wash with 20 mM imidazole/PBS until baseline UV absorbance is reached. Note that imidazole absorbs at 280 nm, so the baseline will be slightly higher than the PBS wash.
5. Elute the protein with 150 mM imidazole in PBS.[a] A typical UV optical recorder profile (OD 280 nm) for the purification of H2-$D^b(His)_6$ is shown in *Figure 4*.
6. Collect and pool the peak fraction.

[a] Ni–agarose columns can be used for the same protein multiple times. Store the used column material at 4 °C in PBS/0.2% sodium azide.

Protocol 6. Anion exchange purification of soluble class I molecules

Equipment and reagents

- Low pressure chromatography system capable of running gradient programs, e.g. Pharmacia FPLC
- UV detector
- Fraction collector
- Anion exchange column—Pharmacia Mono Q column
- 50 mM Tris–HCl pH 8.4
- 50 mM Tris–HCl pH 8.4 containing 1 M NaCl

Method

1. Dialyse protein from *Protocol 5*, step 6 against 100 vol. of 50 mM Tris–HCl pH 8.4 for at least 8 h at 4 °C.
2. Filter protein solution through a 0.22 μm filter.

3. Load the protein on to the anion exchange column at 1 ml/min.
4. Elute the protein with a 0–200 mM NaCl gradient in 20 min. The class I protein is normally eluted by 80–120 mM NaCl. A typical UV optical recorder profile (OD 280 nm) for the purification of H2-D^b $(His)_6$ is shown in *Figure 5*.
5. Analyse fractions by SDS–PAGE (see *Figure 5*) and pool these containing pure class I.
6. Dialyse pooled fraction against 1000 vol. of buffer of choice, e.g. crystallization buffer or PBS.
7. Concentrate pooled fractions using a Centricon-30 (Amicon) to 1–15 mg/ml and store concentrate at 4 °C in the presence of 0.02% sodium azide. Purified empty class I preparations are relatively stable at 4 °C with little noticeable deterioration over six months.

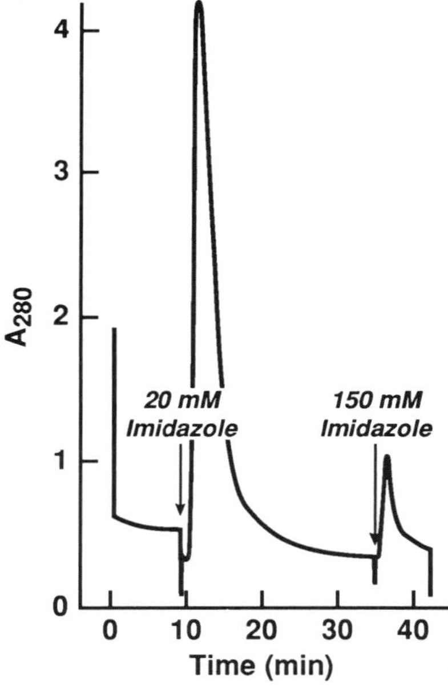

Figure 4. Chromatogram of Ni–agarose column purification of D^bsol $(His)_6$.

6. Measurement of equilibrium binding constant for soluble class I–peptide complexes

Equilibrium binding constants for class I–peptide complexes can be determined by direct binding or by a competition binding assay (9, 18, 19). We prefer the latter approach as it allows the determination of affinities for any

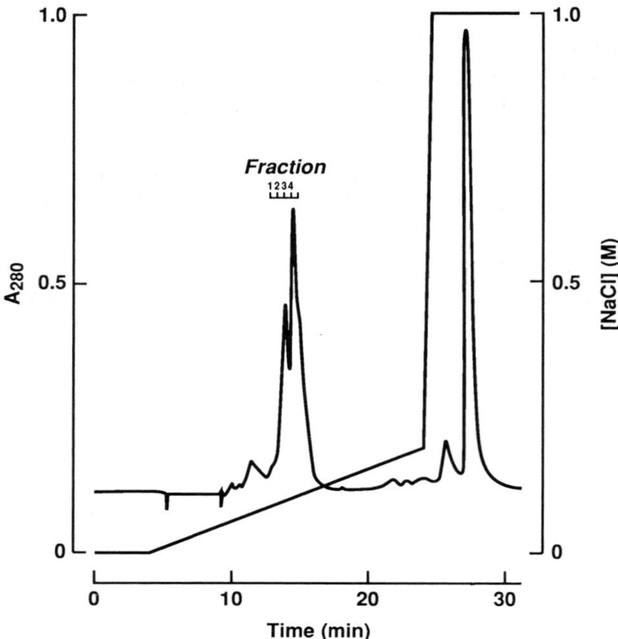

Figure 5. Chromatogram of Mono Q anion exchange purification of protein eluted from Ni–agarose purification step.

peptide of your choice, whereas the former approach is limited to the peptide you have labelled. Several methods have been used for peptide labelling, e.g. radiolabelling with ^{125}I or ^{3}H, or labelling with fluorescent dyes. For direct binding assays you have to make sure that the labelling method produces a peptide with uncompromised ability to bind to the class I (20, 21).

The competition binding assay used in our laboratory requires radiolabelling of a tyrosine or histidine residue, contained in the peptide which is used as a tracer (9, 18). This tracer is then used for competitive binding assays with any peptide of your choice. The mathematical background for the method is given in ref. 19. The technique has been used for determination of dissociation constants (K_D) for the binding of several peptides to K^b, L^d, and D^b. K_D values in the range 10^{-4} to 10^{-10} have been measured using this method. For these studies the following peptides were used as tracer peptides:

K^b VSV-8 (RGYVYQGL)
Ld MCMV (YPHFMPTNL)
D^b SEV-9 (FAPGNYPAL).

Prior to iodination it is essential that the tracer peptide is purified to homogeneity by C18 reverse-phase HPLC.

6.1 Radiolabelling of tracer peptide

Note: labelling procedures must be carried out in a fume-hood certified for ^{125}I work and all workers should be appropriately trained in the use of radioisotopes. γ-Irradiation is of high energy so make sure you have a good supply of lead bricks to protect yourself. We describe two protocols (22). We have found that the former worked well for SEV-9 and MCMV peptide whereas the latter works better for VSV-8 peptide. The reason for this is unclear and the reader is advised to determine empirically which method to use for other peptides. Another potentially useful method for peptide labelling not described here is labelling using IODO-BEADS (Pierce) (23).

Protocol 7. Labelling by lactoperoxidase method

Equipment and reagents

- 2 mM peptide solution in PBS
- Lactoperoxidase (Sigma)
- Glucose oxidase (Sigma)
- 400 mM glucose in PBS
- 10 mM tyrosine in PBS
- ^{125}I, Na salt solution, 100 μCi/μl (Amersham, IMS.30)
- 0.1% trifluoroacetic acid (TFA) in H_2O
- 5% acetonitrile/0.1% TFA in H_2O
- 42% acetonitrile/0.1% TFA in H_2O
- 100% acetonitrile/0.1% TFA
- C18 cartridge (Applied Biosystems, 400771)
- 3 × 10 ml syringes, 1 × 1 ml syringe (syringes should have luer-lock fittings)

Method

1. Dilute lactoperoxidase to 2 U/ml with PBS.
2. Dilute glucose oxidase to 4 U/ml with PBS, keep on ice.
3. To a 1.5 ml Eppendorf tube add the following: 10 μl 2 mM peptide, 10 μl 400 mM glucose, 5 μl lactoperoxidase.
4. Add 0.5 mCi Na ^{125}I.
5. Add 5 μl glucose oxidase and incubate 10 min.
6. Meanwhile equilibrate the C18 cartridges by injecting 10 ml 100% acetonitrile/0.1% TFA, followed by an injection of H_2O/0.1% TFA.
7. Add 25 μl 10 mM tyrosine to reaction mixture from step 5, incubate 2 min.
8. Add 1 ml H_2O/0.1% TFA. Aspirate with 1 ml syringe (most easily done by fitting a cut 5–200 μl pipette tip on to the syringe). Inject labelling mixture on C18 cartridge, collect flow-through.
9. Wash by injecting 5 ml 0.1% TFA, collect flow-through in same tube as step 8.
10. Wash by injecting 3 ml 5% acetonitrile/0.1% TFA, collect flow-through.
11. Elute with 2 ml 42% acetonitrile/0.1% TFA.
12. Measure radioactivity in flow-through from steps 8 and 9, 10, and

Protocol 7. *Continued*

eluted fraction. Most of the unincorporated ^{125}I is in the flow-through from steps 8 and 9. The flow-through from step 10 should contain less than 5% of the total radioactivity. The eluate normally contains over 15% of the total activity.

13. Dry eluate from step 12 in SpeedVac or similar device.

Note: at this stage all iodine should be bound and can be handled outside the fume-hood. However, it is recommended that dried ^{125}I-labelled peptide be handled in a dedicated fume-hood.

Protocol 8. Labelling by chloramine T method

Equipment and reagents

- 1 M $NaPO_4$ pH 7.5
- Stop buffer: 2.4 mg/ml Na metabisulfite in saturated aqueous tyrosine solution (10 mg/ml)
- 2 mg/ml chloramine T
- All items listed in *Protocol 7* except glucose oxidase, lactoperoxidase, and glucose

Method

1. To a 1.5 ml Eppendorf tube add 12.5 μl 2 mM peptide and 12.5 μl $NaPO_4$.
2. Add 0.5 mCi $Na^{125}I$.
3. Add 25 μl chloramine T and incubate 8 min.
4. Meanwhile equilibrate the C18 cartridges by injecting 10 ml 100% acetonitrile/0.1% TFA, followed by an injection of H_2O/0.1%TFA.
5. Add 50 μl stop buffer, incubate for 2 min.
6. Follow *Protocol 7*, step 8 onwards.

Protocol 9. HPLC purification of ^{125}I-labelled peptide

The introduction of an iodine atom in the peptide makes the peptide more hydrophobic and it is therefore possible to separate from unlabelled peptide by reverse-phase chromatography.

Equipment and reagents

- Mobile phase A: H_2O/0.1% TFA
- Mobile phase B: 100% acetonitrile/0.1% TFA
- HPLC set-up capable of running gradient programs
- C18 HPLC column
- UV detector
- Fraction collector
- γ-Radiation absorbing transparent shields

Method

1. Add 190 μl H_2O/0.1% TFA to 10 μl of the purified peptide solution used in *Protocol 7*. Set your HPLC to run a gradient from 5% B to 30% B in 45 min, followed by 10 min of 100% B and 10 min of 5% B.
2. Inject solution from step 1 and do a test run to check where the peptide elutes in your gradient.
3. Resuspend dried radiolabelled peptide from *Protocol 7*, step 13 in 200 μl H_2O/0.1% TFA.
4. Inject peptide and run gradient program. Collect fractions with volumes around 300 μl.
5. Measure radioactivity in each fraction. The iodinated peptide elutes after the unlabelled peptide. Most of the radioactivity should be in one or two fractions. It is very important that the unlabelled and labelled fractions are well separated from each other. Dry down the one or two fractions containing most of the radioactivity in a SpeedVac or similar device.
6. Resuspend in 1 ml PBS. Measure total radioactivity.

Note: to determine the binding of high affinity peptides to class I molecules it is important to know the concentration of your tracer peptide (see below). The concentration is difficult to measure accurately, e.g. using a spectrophotometer. Instead we usually calculate the concentration of the tracer peptide from the specific activity of the iodine batch used for labelling using the following equation:

$$\text{Concentration (molar)} = T_a/(S_a \times 126.9 \times \text{Vol.})$$

T_a = total activity in Ci, S_a = specific activity in Ci/g, and Vol. = volume in litres. Remember to compensate for decay of your iodine preparation. The half-life of ^{125}I is 60 days.

Protocol 10. Calibration of gel filtration column

We use Bio-Rad P-30 gel filtration spin columns, but any gel filtration column with similar bed volume and exclusion limit should work.

Equipment and reagents

- Two gel filtration columns
- 100 μl ^{125}I-labelled peptide (20 000 c.p.m.) in PBS with 1% BSA and 0.02% sodium azide
- Test-tubes: 4 ml plastic tubes
- Spectrophotometer
- 100 μl of a protein solution with an absorbance at 280 nm of approx. 1 (the protein should have a M_r not higher than 45 kDa)
- γ-counter

Protocol 10. *Continued*

Method

1. Wash the columns with 4 ml PBS.
2. Put the columns over a test-tube so that the eluate can be collected.
3. Load 95 μl of protein on one of the columns and 95 μl of ^{125}I-labelled peptide on the other.
4. Wait until the solution has entered the column.
5. Move the column to a second test-tube.
6. Add 100 μl PBS.
7. Repeat steps 4–6 until a total of 1 ml of PBS has been loaded on to the column.
8. Measure A_{280} of the protein column and radiation of the peptide fractions on a γ-counter. Also measure remaining radioactivity on the column.
9. Plot the percentage of total protein and radioactivity eluted in each fraction to determine optimal separation.

Protocol 11. Measurement of half-maximal binding of tracer peptide to class I molecule

Equipment and reagents

- Gel filtration columns
- PBS with 1% BSA
- Siliconized Eppendorf tubes
- Test tubes: 4 ml plastic
- Tracer peptide from *Protocol 9*, step 6
- Purified empty soluble class I (1–5 mg/ml in PBS) *Protocol 6*, step 7
- γ-counter

Method

1. Make a dilution series of your class I molecule from 10 μM down to 500 pM.
2. Prepare ten Eppendorf tubes containing 20 000 c.p.m. tracer peptide in 90 μl 1% BSA/PBS.
3. Add 10 μl of the class I protein dilutions from step 1 to the ten Eppendorf tubes prepared in step 2, mix, and spin down.
4. Incubate at room temperature until equilibrium has been reached, 3 h or K^b. Note: 3 h is sufficient for the K^b ^{125}I-VSV-8 complex but the time required to reach equilibrium varies between different class I–tracer peptide complexes, e.g. the L^d ^{125}I-MCMV complex requires 24 h to each equilibrium.
5. 1 h before equilibrium has been reached start washing the gel filtration columns with several volumes of PBS.

6. Add 95 μl of the samples from step 3 to the gel filtration columns.
7. Carefully add the volume of PBS required to elute the class I–peptide complexes (determined in *Protocol 10*).
8. Measure radioactivity in eluted fraction (containing class I–tracer peptide complexes) and columns (containing free tracer peptide) in a γ-counter.
9. Wash the columns with several volumes of PBS.
10. Calculate the fraction of peptide bound at each class I concentration: c.p.m. eluted fraction/(c.p.m. eluted fraction + c.p.m. column). Plot the fraction of peptide bound against class I concentration. Determine the protein concentration that gives 50% binding of peptide. An example of the determination of half-maximal binding of ^{125}I-VSV by K^b is shown in *Figure 6*.

$[K^b]$ (M)	c.p.m./eluted	c.p.m./column	% Binding
$1.0*10^{-5}$	17723	2758	86.5
$3.0*10^{-6}$	17017	2814	85.8
$9.0*10^{-7}$	16559	3574	82.2
$2.7*10^{-7}$	15589	4331	78.3
$8.1*10^{-8}$	12392	7512	62.3
$2.4*10^{-8}$	8416	11638	42.0
$7.3*10^{-9}$	4530	15478	22.6
$2.2*10^{-9}$	2292	17835	11.4
$6.7*10^{-10}$	896	19076	4.49
$6.7*10^{-11}$	590	19612	2.92

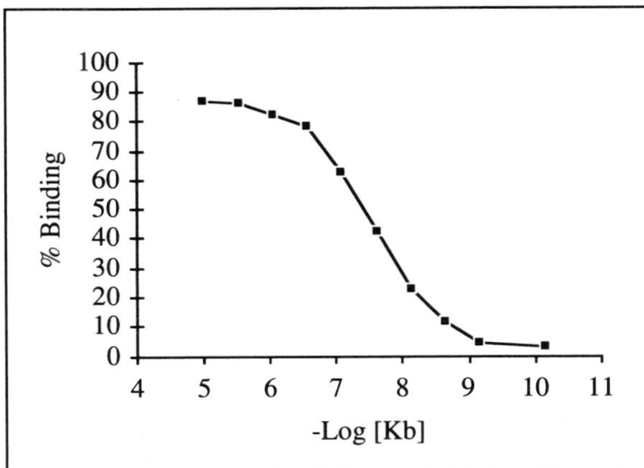

Figure 6. Determination of half-maximal binding of ^{125}I–VSV peptide to K^b. Values were obtained as described in *Protocol 11*. Half-maximal binding of tracer occurs at a K^b concentration of 36 nM.

6.2 Determination of equilibrium binding constants for peptide binding to class I

In this protocol we describe how to determine equilibrium binding constants to class I for a particular peptide by incubating the amount of class I needed to achieve half-maximal binding of tracer (*Protocol 11*) with different concentrations of the peptide under study (called inhibitor—I). A good range of inhibitor concentrations to start with is 10^{-5} to 10^{-9} M.

Protocol 12. Determination of equilibrium binding constants

Equipment and reagents

- Solution of peptide(s) under study (I) (100 μM/PBS)
- All items listed in *Protocol 11*

Method

1. Prepare a set of Eppendorf tubes containing 20 000 c.p.m. tracer peptide, the concentration of class I giving 50% binding of tracer peptide, 10 μl of the peptide under study (I) at different concentrations. Make up to 100 μl with 1% BSA/PBS. Include a tube without inhibitor for determination of maximal binding ($M_{1/2}$) and one tube with only tracer for determination of background binding (Bg).
2. Incubate for the time required to reach equilibrium.
3. Separate free and class I-bound tracer peptide as described in *Protocol 11,* and calculate the fraction tracer bound at each inhibitor concentration as described in *Protocol 11*. Calculate the inhibition of tracer binding to class I at each concentration of inhibitor by the equation:
$$\% \text{ inhibition} = 100 \times ((M_{1/2} - Bg) - (B - Bg))/(M_{1/2} - Bg)$$
where B is the fraction tracer bounds at the different inhibitor concentrations used.
4. Plot per cent inhibition against [I]. Determine the concentration which gives 50% inhibition. The dissociation constant is then given by:
$$K_D = 3/8 \times ([I_t] - [T_t])$$
where $[I_t]$ is the concentration of inhibitor giving 50% inhibition and $[T_t]$ is the concentration of tracer in the reaction mixture. If Bg is high (over 5–10%) there may be a problem with your labelling or purification of tracer peptide.

References

1. Bjorkman, P. J., Saper, M. A., Samraoui, B., Bennett, W. S., Strominger, J. L., and Wiley, D. C. (1987). *Nature*, **329**, 506.

2. Ljunggren, H.-G., Stam, N. J., Öhlen, C., Neefjees, J. J., Höglund, P., Heemels, M.-T., *et al.* (1990). *Nature*, **346**, 476.
3. Fahnestock, M. L., Tamir, I., Narhi, L., and Bjorkman, P. M. (1992). *Science*, **258**, 1658.
4. Silver, M. L., Parker, K. C., and Wiley, D. C. (1991). *Nature*, **350**, 619.
5. Zhang, W., Young, A. C. M., Imari, M., and Nathenson, S. G. (1992). *Proc. Natl. Acad. Sci. USA*, **89**, 8403.
6. Parker, K. C., Carreno, B. M., Sestak, B. M., Utz, U., Biddison, W. E., and Coligan, J. E. (1992). *J. Biol. Chem.*, **267**, 5451.
7. Lévy, F. and Kvist, S. (1990). *Int. Immunol.*, **2**, 996.
8. Jackson, M. R., Song, E. S., Yang, Y., and Peterson, P. A. (1992). *Proc. Natl. Acad. Sci. USA*, **89**, 12117.
9. Matsamura, M., Saito, Y., Jackson, M. R., Song, E. S., and Peterson, P. A. (1992). *J. Biol. Chem.*, **267**, 23589.
10. Jackson, M. R., Cohen-Doyle, M. F., Peterson, P. A., and Williams, D. B. (1994). *Science*, **263**, 384.
11. Bunch, T. A., Grinblat, Y., and Goldstein, L. S. B. (1988). *Nucleic Acids Res.*, **6**, 1043.
12. Kozak, M. (1993). *Microbiol. Rev.*, **47**, 1.
13. Fields, J., Nikawa, J., Broek, D., MacDonald, B., Rodgers, L., Wilson, I. A. *et al.* (1988). *J. Mol. Biol.*, **8**, 2159.
14. Evans, G. I., Lewis, G. K., Ramsay, G., and Bishop, J. M. (1985). *Mol. Cell. Biol.*, **5**, 3610.
15. Hoffman, A. (1990). In *Genetic engineering, principle and methods* (ed. J. K. Setlow) pp. 87–98. Plenum Press, New York.
16. Ausubel, F. M., Brent, R., Kingston, R. E., Moore, D. D., Smith, J. A., Seidman, J. G. *et al.* (ed.) (1988). Short protocols in molecular biology, third edition. Greene Publishing Associates and John Wiley and Sons, New York.
17. Steller, H. and Pirrotta, V. (1985). *EMBO J.*, **4**, 167.
18. Saito, Y., Peterson, P. A., and Matsumura, M. (1993). *J. Biol. Chem.*, **268**, 21309.
19. Müller, R. (1983). In *Methods in enzymology* (ed. J. J. Langone and H. van Vunakis) Vol. 92, pp. 589–601. Academic Press, New York.
20. Cerundolo, V., Elliot, T., Elvin, J., Bastin, J., Rammensee, H. G., and Townsend, A. (1991). *Eur. J. Immunol.*, **21**, 2069.
21. Corr, M., Boyd, L. F., Frankel, S. R., Kozlowski, S., Padlan, E. A., and Margulies, D. H. (1992). *J. Exp. Med.*, **176**, 1681.
22. Harlow, E. and Lane, D. (1988). *Antibodies: a laboratory manual.* Cold Spring Harbor Press, New York.
23. Tsomides, T. J. and Eisen, H. N. (1993). *Anal. Biochem.*, **210**, 129.

5

The use of pulsed-field gel electrophoresis to map the human MHC

PHILIPPE SANSÉAU and JOHN TROWSDALE

1. Introduction

The major histocompatibility complex (MHC) is located on the short arm of chromosome 6 in the distal portion of the 6p21.3 band. It consists of three linked gene clusters covering 4 Mb of DNA. The class I and class II regions encode highly polymorphic families of cell surface glycoproteins involved in immune regulation. The class I region which covers ~1500 kb, contains at least 40 genes including the classical transplantation antigens (HLA-A, -B, and -C). This region is located at the telomeric side of the MHC. The class II region encodes the classical antigens (HLA-DP, -DQ, -DR) but also novel class II genes like HLA-DM, and the TAPs and LMPs involved in antigen processing. The class I and class II regions are separated by the class III region which spans ~ 1100 kb. This part of the MHC contains a large number of genes unrelated to the classical transplantation antigens, although some are involved in the immune system, such as the complement genes C2, C4, and Bf, the tumour necrosis factor (TNF) A and B genes, and the heat shock protein HSP70 genes.

Conventional agarose gel electrophoresis of DNA is limited to the separation of DNA up to 50 kb in size. A different type of electrophoresis called pulsed-field gel electrophoresis (PFGE) (1–6), capable of separating DNA molecules up to 10 Mb, can be used to produce long-range restriction maps by hybridization of various probes on genomic Southern blots (7–10). Rare cutter enzymes are often used in PFGE since they produce large fragments and are useful to identify the HTF (*Hpa*II tiny fragment) islands (11). These are CpG-rich, unmethylated stretches of DNA and are often associated with the 5′ ends of genes. Once an island has been found, probes may be used to search for associated transcripts. PFGE is a powerful technique essential to characterize a large genomic region such as the MHC.

2. Pulsed-field gel electrophoresis

2.1 Types of apparatus

PFGE was first described by Schwartz and Cantor in 1984 (1). In this method the DNA is subjected to two electric fields. The angle between these is approximately perpendicular, the fields are non-uniform in strength, and are alternatively switched and pulsed (see *Figure 1a*). Under these conditions the DNA molecules need to reorient themselves in the gel at each inversion of the electric field. They move in a direction approximately at right angles to the axis along which they are stretched. The time taken to find the new orientation is a function of the size of the DNA molecule. A long molecule is held back in the gel in comparison with a short one. Under these conditions Schwartz and Cantor were able to separate yeast chromosomal DNAs up to an estimated size of 2 Mb.

A number of modifications of the original method have followed rapidly to improve the resolution and to produce linear DNA trajectories. These modifications concern the arrangement of electrodes, the angles between the fields, the switching regimes, and types of field. Carle and Olson (2) employed a double inhomogeneous field to produce orthogonal field alternation gel electrophoresis (OFAGE) (see *Figure 1b*). In this configuration, the technique resolves DNA into straight tracks, as in homogeneous field electrophoresis. In 1986 Carle and Olson (3) used a conventional horizontal gel apparatus and the polarity of the field was switched at each cycle. This method of PFGE is called field inversion gel electrophoresis (FIGE) (see *Figure 1c*). The net forward movement of the DNA molecules is achieved by using a greater switching time for forward than for reverse migration or a higher voltage in the forward than in the reverse direction. In this system the DNA migrated along linear trajectories but the relationship between size and mobility is not well understood. To overcome this problem the homogeneous crossed field electrophoresis was developed. In this type of apparatus the DNA migrates in straight tracks and the mobility of the DNA molecules is a function of their sizes for most of the gel. The first apparatus was a circular gel in a square gel box (4) (see *Figure 1d*). At each switching time the gel is turned, therefore the DNA experiences a change in the field angle greater than 90°. A similar system has also been described by Serwer (5). The second form of homogeneous field apparatus has been described by Chu *et al.* (6). In this system the electrodes are connected by a series of resistors (see *Figure 1e*) and form a hexagon. The electric fields are at 120° to one another and the electrodes are individually voltage clamped or autonomously controlled to provide a near homogeneous field. Therefore the technique has been named contour clamped homogeneous electric (CHEF) gel electrophoresis. The CHEF method has been used successfully to separate DNA molecules up to 10 Mb (12).

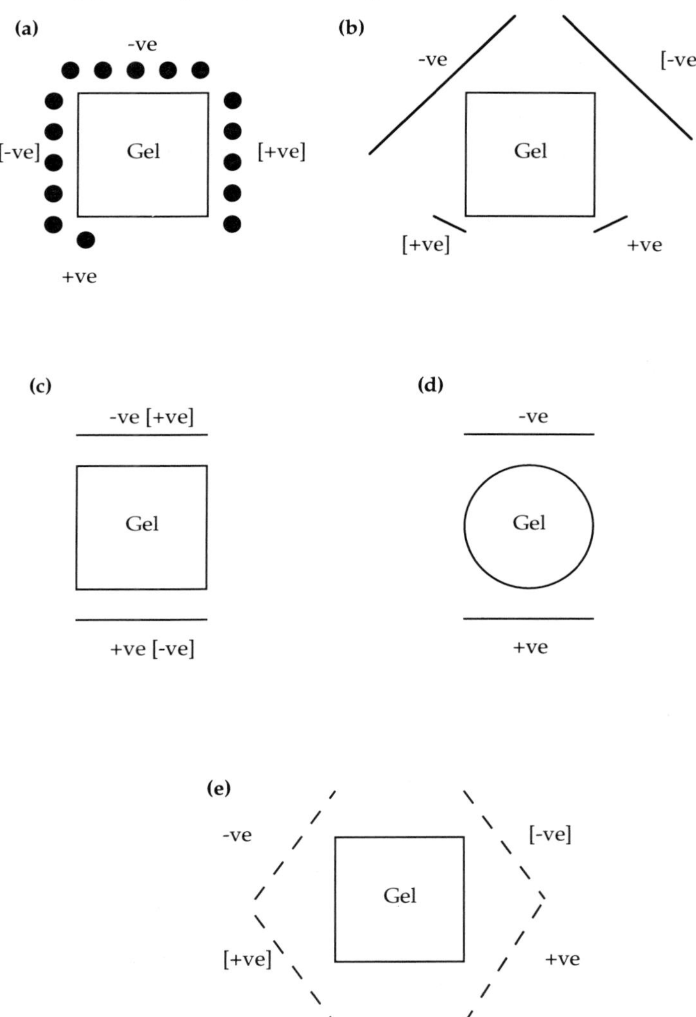

Figure 1. Electrode configurations used in the different PFGE apparatus. (a) First system described by Schwartz and Cantor (1) in 1984. The dots represent the diode isolated electrodes. (b) Orthogonal field alternation gel electrophoresis (OFAGE) (2). (c) Field inversion gel electrophoresis (FIGE) (3). (d) Rotating pulsed-field system (4, 5). (e) Contour clamped homogeneous electric field (CHEF) (6).

2.2 Applications of the PFGE

This technique has many applications. First, in lower eukaryotes (i.e. yeast or trypanosomes and mycoplasma) PFGE is able to separate individual chromosomes to establish karyotypes. It is also possible to purify the chromosomes in order to construct chromosome-specific libraries. In higher organisms and in

Table 1. Rare cutter restriction endonucleases used for PFGE and their predicted average fragment size in mammalian DNA

Enzyme	Recognition sequence[a]	Average fragment size (in kb)
*Bss*HII	GCGCGC	100
*Cla*I	ATCGAT	100
*Eag*I (*Xma*III)	CGGCCG	100
*Ksp*I	CCGCGG	100
*Mlu*I	ACGCGT	300
*Nae*I	GCCGGC	100
*Nar*I	CGCGCC	100
*Not*I	GCGGCCCGC	1000
*Nru*I	TCGCGA	300
*Pvu*I	CGATCG	300
*Sal*I	GTCGAC	100
*Sfi*I	GGCCNNNNNGGCC	200
*Sma*I	CCCGGG	100
*Xho*I	CTCGAG	100

[a] N: any nucleotide.

mammals such as humans, chromosomes have not been yet separated as individual bands. The smallest human chromosome, 21, is estimated to be about 45 Mb in size.

The second main application of the technique is the construction of long-range physical maps from large genomic regions. For mapping mammalian DNA which is much bigger than the yeast genome, it is necessary to cut the genomic DNA into fragments that can be resolved by PFGE. There are commercially available restriction endonucleases known to cleave the DNA relatively rarely to produce fragments in the 50–1000 kb range (see *Table 1*). These enzymes can have a long recognition sequence, for example *Sfi*I or *Not*I. The most useful class of endonucleases are those which are methylation-sensitive and contain one or more CpG dinucleotides in their recognition sites. CpG is relatively rare in the mammalian genome (11) and 70–90% of these dinucleotides are methylated. Analysis of DNA fragments by blotting and hybridization of probes to common DNA on Southern blots allow the establishment of physical maps. If YACs (yeast artificial chromosomes) are used to construct maps the restriction sites are demethylated in comparison with the equivalent genomic DNA. Moreover the degree of methylation could be variable from one cell line from another and this effect obscures the mapping.

3. Preparation of samples

For this type of electrophoresis very large size DNA is used, therefore it is essential to avoid any nuclease contamination. When possible, all solutions,

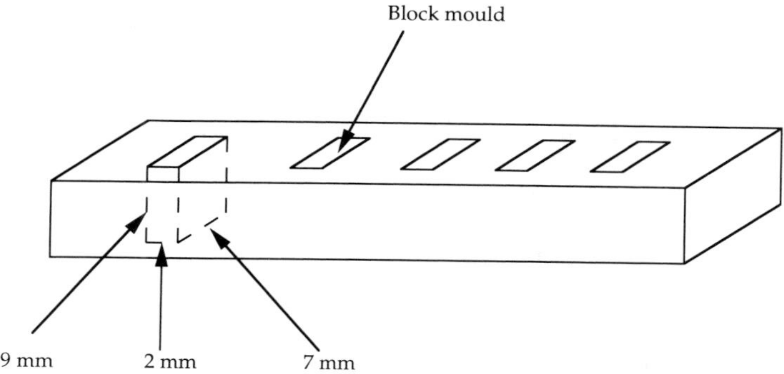

Figure 2. A Perspex mould used for preparing DNA embedded in agarose.

tubes, and tips should be autoclaved. Solutions which cannot be autoclaved should be made up in glass double distilled, double deionized autoclaved water, and then filtered through a 0.22 μm sterile filter unit. Gloves should be worn during all manipulations.

3.1 Plug mould

To prepare high molecular weight DNA embedded in blocks you need to pour the agarose into a plug mould. The authors use a Perspex plug mould which gives plugs of ~ 7 × 2 × 9 mm in size (see *Figure 2*). The size of the plugs is not important as long as the agarose blocks match the size of the wells of the gel.

3.2 Solutions

You will need the following solutions:

(a) YPD medium: this medium is used to grow yeast strains. Make a 2% glucose, 2% bactopeptone, and 1% yeast extract solution.

(b) Low gelling temperature (LGT) agarose: the authors use NuSieve GTG agarose (FMC Bioproducts).

(c) L buffer: make a solution of 0.1 M EDTA, 0.01 M Tris–HCl pH 7.6, 0.02 M NaCl.

(d) NDS: 1% Sarkosyl, 0.5 M EDTA, 0.01 M Tris–Hcl pH 9.5. This solution is used in the preparation of DNA embedded in agarose plugs.

(e) Proteinase K (BDH laboratories): dissolve in H_2O at a concentration of 20 mg/ml, dispense in 500 μl aliquots, and store at −20°C.

(f) Lyticase (Sigma Chemical Co): dissolve in 0.01 M sodium phosphate containing 50% glycerol to a concentration of 90 U/μl and store at −20°C.

(g) Phenylmethylsulfonyl fluoride (PMSF) (Sigma Chemical Co): PMSF is extremely toxic and should be handled with great care. Make a 10 mM or 100 mM stock solution (1.74 or 17.4 mg/ml) in isopropanol and store at −200°C.

3.3 Markers for PFGE

Markers used for PFGE are usually yeast chromosomes or lambda concatemers. For lower size range it is possible to use a *Hin*dIII digest of lambda DNA. *Table 2* describes these different markers.

3.3.1 Yeast chromosome markers

Saccharomyces cerevisiae strains are used in the range 100 kb to 2000 kb (see *Figure 3*) and chromosomes from *Schizosaccharomyces pombe* are useful at the limits of PFGE resolution. The following protocol is used by the authors to prepare intact DNA from yeast.

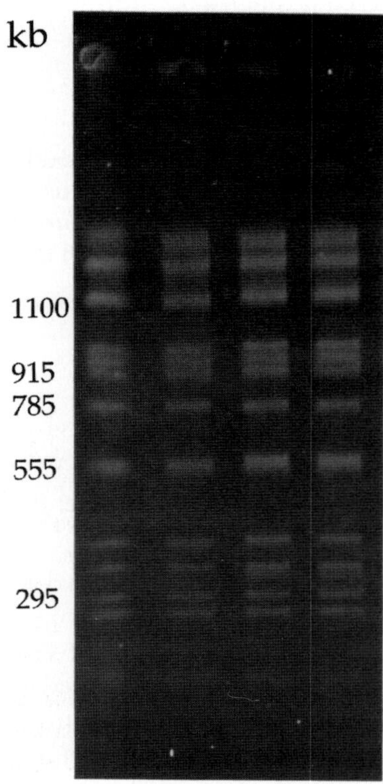

Figure 3. PFGE separation of yeast chromosomes (strain AB1380)> PFGE (CHEF) running conditions were 200 V at 10°C for 12 h at 6 sec pulses, followed by 90 sec pulses for 10 h on a 1% agarose gel in 0.5 × TBE buffer.

Table 2. Molecular weight markers commonly used for PFGE (in kb)

λ-concatemers	***Saccharomyces cerivisiae* chromosomes (strain YPH 501)[a]**
873	2200
824.5	1650
776	1100
727.5	1005
679	960
630.5	833
582	802
533.5	737
485	649
436.5	530
388	497
339.5	419
291	303
242.5	260
194	222
145.5	
97	
48.5	

[a] The sizes vary between different yeast strains.

Protocol 1. Preparation of chromosomal DNA markers from *Saccharomyces cerevisiae*

Reagents

- See Section 3.2

Method

1. Streak cells on to a YPD–agar plate (2% agar in YPD) and grow at 30 °C.
2. Pick one single colony and grow in 30 ml of YPD for two days.
3. Collect yeast cells growing in suspension by centrifugation at 3000 *g* for 5 min at 4 °C. Remove the supernatant and wash the pellet twice with 0.05 M EDTA pH 8, 0.01 M Tris–HCl pH 7.6.
4. Resuspend the cells at a concentration of 3×10^{10} cells/ml in 0.05 M EDTA pH 8.
5. Dissolve LGT agarose to a concentration of 1% in L buffer. Cool the melted agarose to 42 °C.
6. Stick tape on to one surface of the clean Perspex plug mould. Place the mould on ice.
7. Warm the cell suspension to 42 °C. Mix equal volumes of LGT agarose and yeast cell suspension. Stir the mixture with a sealed Pasteur pipette.

Protocol 1. *Continued*

8. Add 75 μl of lyticase and mix. Before the agarose sets, pipette the molten mixture into the plug mould. Allow to set on ice for 15 min.
9. Collect the blocks and incubate for 24 h at 37 °C in 50 vol. of 0.5 M EDTA pH 8, 0.01 M Tris–HCl pH 7.6 containing 1% β-mercaptoethanol.
10. Transfer the blocks to 50 vol. of NDS containing 1 mg/ml proteinase K. Incubate the blocks 24 h at 50 °C. Replace the mixture with a fresh solution and incubate for a further 24 h.
11. Wash the blocks twice in NDS and store at 4 °C in the same solution.
12. Yeast chromosomes are stable in the blocks for several months at 4 °C.

3.3.2 Concatemers of lambda DNA

At the lower size range and up to 1000 kb lambda concatemers are appropriate standards for PFGE. These concatemers yield a series of markers which vary in size by one molecule (i.e. 48.5–~ 1000 kb) (see *Figure 4a*). The following protocol describes an easy and inexpensive method to prepare lambda concatemers.

Protocol 2. Preparation of concatemers of lambda

1. Prepare phage lambda DNA as described in ref. 13. Concentrate the DNA by using butanol extraction to have a final concentration of approx. 0.25 mg/ml. Use gentle pipetting to avoid any shearing.
2. Incubate the phage DNA at a concentration of 200 μg/ml in 2 × SSC containing 3% Ficoll and a little orange G as dye for 30 min at 37 °C. Place the solution on the bench at room temperature overnight.
3. Store the concatemers at 4 °C. They are stable for several months.

3.4 Isolation of DNA for PFGE from cultured cells

For PFGE mammalian DNA should be embedded in agarose to avoid any shearing. The following protocol describes a method used for cultured cells.

Protocol 3. Preparation of DNA plugs from cultured cells

Reagents

- See Section 3.2

Method

1. Wash the cultured cells in PBS twice and resuspend in PBS at a concentration of 2×10^7 cells/ml.

2. Make 2% LGT agarose in PBS and hold at 42 °C.
3. Stick tape on to one side of the Perspex plug mould and place on ice.
4. Place the cells at 42 °C, mix with an equal volume of LGT agarose, and dispense into the plug mould.
5. Allow blocks to set on ice for 15 min.
6. Remove the blocks and incubate at 50 °C in 50 vol. of NDS solution containing 1 mg/ml of proteinase K for 24 h. Remove the solution, add 10 ml of the same fresh NDS–proteinase K solution, and incubate at 50 °C for a further 48 h.
7. Wash blocks twice in NDS and store in the same solution at 4 °C. The blocks are stable for several months.

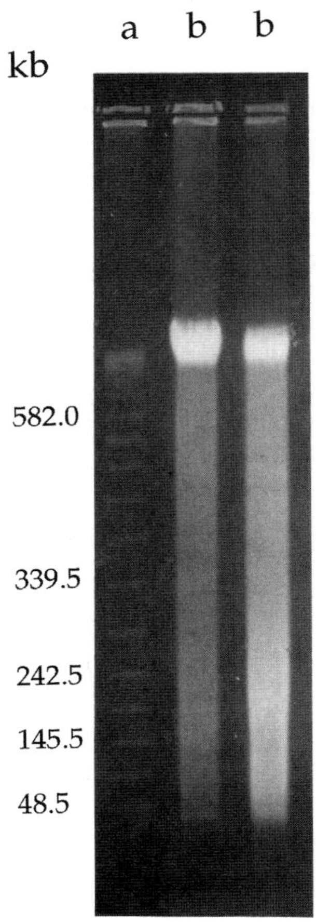

Figure 4. PFGE separations of (a) lambda concatemers and (b) cell line DNAs digested with *Nru*I. PFGE (CHEF) running conditions were 200 V at 10 °C for 24 h with 50–90 sec pulses ramp on a 1% agarose gel in 0.5 × TBE buffer.

3.5 Restriction enzyme digests of DNA in plugs

The following protocol can be used to digest mammalian DNA or yeast DNA such as YAC clones (see *Figure 4b*).

3.5.1 Total digests

Protocol 4. Restriction enzyme digestion of DNA embedded in agarose plugs

Reagents

- See Section 3.2

Method

1. Cut a plug containing 5–10 μg of DNA into three parts for each digest.
2. Wash the blocks from a same cell line three times in 50 vol. of TE (10 mM Tris–HCl, 1 mM EDTA pH 8) containing 40 μg/ml of PMSF to inactivate the proteinase K. Rinse the plugs three times in TE.
3. Equilibrate each plug with 500 μl of the appropriate restriction enzyme buffer in a microcentrifuge tube for 30 min. Use the restriction enzyme buffer specific for the enzyme without gelatin (or BSA), DTT (or β-mercaptoethanol), or spermidine.
4. Replace the solution with 100 μl of fresh restriction buffer containing 500 μg/ml BSA, 1 mM DTT, and spermidine. Add the restriction enzyme, mix, and incubate at the appropriate temperature for 2–3 h.
5. For double digestion, after the first digest equilibrate the blocks as in step 3 with the appropriate restriction enzyme buffer. Add the new enzyme as in step 4.
6. After incubation place the blocks at 4°C for 15 min and load into the PFGE gel. Seal the samples with 0.5% LGT agarose. Digested plugs can be stored for six months in NDS.

3.5.2 Partial digests

When the DNA is in solution partial digests are achieved by reducing the enzyme concentration or the incubation time. Since the DNA is embedded in agarose for PFGE a combination of both is often used. For example, each plug can be pre-incubated with three different enzymes concentrations. After diffusion of the enzymes through the blocks, incubate for two different time points.

4. Electrophoresis conditions

Conditions for PFGE vary in function of the apparatus being used and the separation range required. Basically, the migration of the DNA fragments is dependent on the gel concentration, the voltage gradient, the temperature, the reorientation angle, and the switching interval. The last parameter is the most useful. By increasing the switching interval, DNA molecules of increasing size are separated. In some cases it is worth using a ramp for the switching time. The reader should refer to the manufacturer's manual for each apparatus to establish the best conditions necessary to separate DNA fragments of a particular size range.

Generally, electrophoresis is carried out in 0.5 × TAE (5 × TAE contains 24.2 g Tris base, 5.71 ml glacial acetic acid, and 20 ml 0.5 M EDTA pH 8 per litre), or 0.5 × TBE (5 × TBE contains 54 g Tris base, 27.5 g boric acids, and 20 ml of 0.5 M EDTA pH 8 per litre), at a constant temperature and with a fixed voltage across the electrodes.

5. Gel processing

After electrophoresis stain the gel in water containing ethidium bromide (0.5 μg/ml) for 30 min–1 h. Destain for up to 1 h in water and then photograph using a UV transilluminator (306 nm).

5.1 Gel blotting

Protocol 5. Blotting on to nylon membranes

1. Depurinate the gel in 0.25 M HCl for 20 min.
2. Denature and neutralize as recommended by supplier.
3. Transfer the DNA on to the nylon membrane by blotting overnight in 20 × SSC. The authors also use the Vacu-blot system from Hybaid.
4. Wash the membrane in 2 × SSC and bake it for 2 h in a vacuum oven at 80 °C for 2 h, or expose to UV light to fix the DNA.

5.2 Probing PFGE blots

Many different hybridization protocols are available. In the next protocol, the authors described one they use successfully.

Protocol 6. Hybridization of PFGE blots

For Hybond N^+ mix concentrated stock solutions to the following final concentrations for pre-hybridization and hybridization at 65 °C.

Reagents

- 6 × SSC
- 5 × Denhardt's solution: 0.1% Ficoll, 0.1% polyvinyl pyrrolidone, 0.1% BSA
- 0.5% SDS
- 10% dextran sulfate
- 100 μg/ml denatured, sonicated salmon sperm DNA

Method

1. Pre-hybridize for at least 3 h at 65 °C.
2. Add the probe and fresh hybridization solution. Hybridize overnight at 65 °C.
3. Wash the filters in 2 × SSC, 0.1% SDS for 20 min at 65 °C (two times), followed by 0.2 × SSC, 0.1% for 20 min at 65 °C (two times).
4. Cover filters in Saran Wrap and expose to X-OMAT Kodak films at −80 °C.

6. Applications of PFGE to mapping the MHC

PFGE was used in order to construct a long-range physical map of the human MHC. The choice of genomic DNA is important since haplotype-specific RFLPs exist. DNA must be from HLA homozygous cell lines to minimize this problem. DNA is cut with rare cutter enzymes (see *Table 1*) and run on to a PFGE gel. After Southern blotting resulting filters can be hybridized with probes located in the MHC. Two probes are likely to be linked if the same restriction DNA fragment hybridizes to both probes. Obviously, double digestions are needed as for mapping in conventional electrophoresis. Partial digests are also useful. By using the results from the PFGE in combination with known genetic data, information of the positions of the probes and restriction sites from clones over the region, a physical map of the MHC can be established.

Class I genes are contained in ~ 2000 kb, class III genes in ~ 1100 kb, and class II genes in ~ 750 kb (for details see ref. 14). The mapping with PFGE has been used to determine the orientation and position of several genes in the MHC and the physical distance between those genes. Moreover PFGE was useful for comparative mapping by using genomic DNA from different populations for analysis (15).

6.1 The class I region

One of the early difficulties in constructing a physical map of the class I region was its large size and available probes. Now hundreds of probes can be used. PFGE analysis has established the relative order and the distance between the class I genes (16–20) (see *Figure 5*). The two centromeric genes, HLA-B and HLA-C are 130 kb apart, while HLA-E lies ~ 700 kb from HLA-C. On the other side, HLA-A lies 600 kb from HLA-E. Three class I genes are located telomeric to HLA-A: HLA-H, HLA-G, HLA-F. The distance between HLA-A and HLA-F is approximately 320 kb. Interestingly it was shown by recombination analysis that HLA-G and HLA-F lie 8 cM telomeric to HLA-A. The difference between the physical and the genetic distances suggests the presence of a recombination hot spot telomeric of HLA-A. Moreover variation in the size of the region has been reported. A distance of 350–490 kb for the region extending from HLA-A to HLA-F has been proposed (21).

With the advent of yeast artificial chromosome (YAC) vectors, it has become possible to clone large regions of genomic DNA within a single fragment. To analyse these clones, PFGE was the method of choice. This technology was applied in cloning all of the class I region and determining the organization of the class I genes. The entire class I region has now been cloned as a series of overlapping YAC clones by several groups (22, 23). These clones can be used to refine the previous published maps and to search for new transcripts.

6.2 The class II region

The first published molecular map of the class II region was established for a DR4 haplotype (7). Interestingly this map showed a similar distance between HLA-DQ to HLA-DR and HLA-DQ to HLA-DP. No recombination between the HLA-DQ and HLA-DR loci has been described and alleles of these genes are in strong linkage disequilibrium. Moreover, recombination (1–3%) takes place between HLA-DP and the HLA-DQ and HLA-DR loci and a weak linkage disequilibrium exists between the HLA-DP and HLA-DR alleles. Since the distances between HLA-DR/HLA-DQ and HLA-DQ/HLA-DP are similar, these data suggest the presence of a recombination hot spot between HLA-DP and the HLA-DQ to HLA-DR region.

Within the class II region is has been shown by PFGE analysis that the size of the HLA-DR to HLA-DQ interval is haplotype-dependent. Moreover the organization and the number of genes in the DR subregion can change from one haplotype to another (24–26). The presence of this additional DNA may be important for the recombination frequency between HLA-DQ and HLA-DR and the maintenance of the linkage disequilibrium in this region. Furthermore, certain subtypes of haplotypes carrying a particular DR specificity are more closely related with autoimmune diseases than others. For example narcoleptic patients are almost exclusively associated with HLA-DR2 DQw1

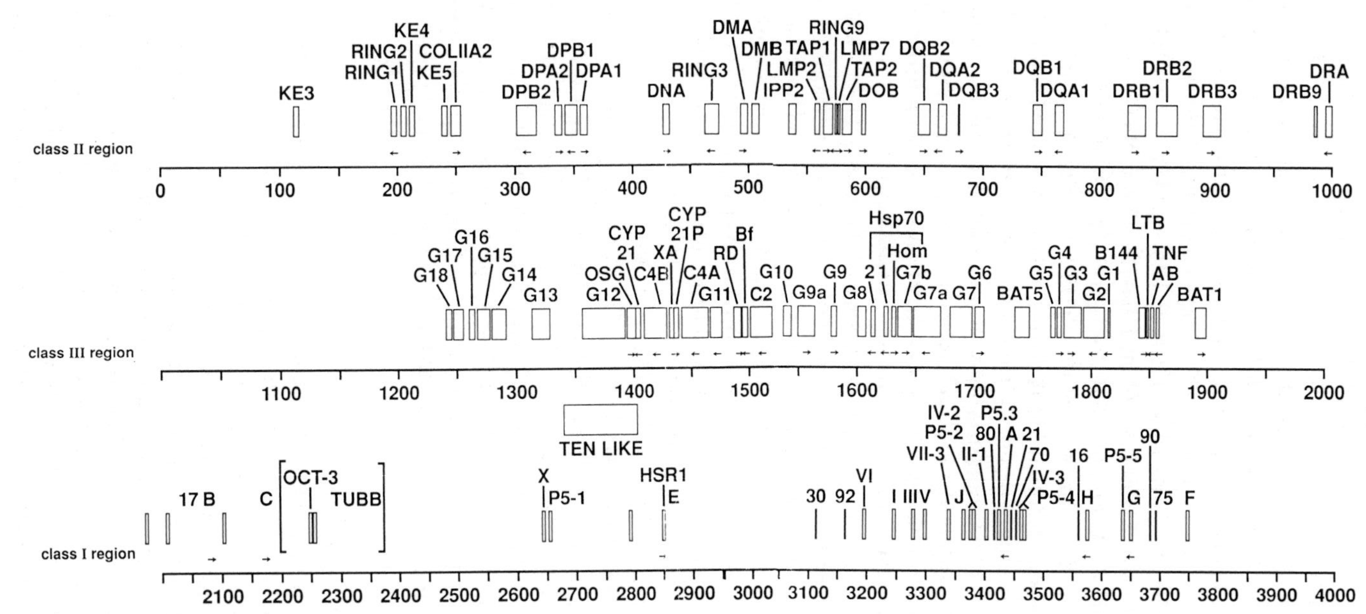

Figure 5. Molecular map of the MHC. This map is a compilation of physical mapping and cloning data from different laboratories.

(27), or insulin-dependent diabetes is associated with DR7 in the black population and within the DR4 Dw4 and DR4 Dw14 haplotypes (28). Different studies have compared a selection of different haplotypes within the class II region. DR4, DR7, and DR9 haplotypes contained between 110–160 kg more DNA than haplotypes carrying the DR3 specificity (24–26, 29). Furthermore, PFGE mapping with YACs derived from two haplotypes has shown the presence of extra DNA in a DR8 haplotype (30, 31). Nevertheless the amount of chromosomal DNA is usually constant for cell lines with the same haplotype. The polymorphic nature of the DQ to DR interval is unusual in comparison with the rest of the class II region. Obviously it will be of interest to look for new coding sequences maybe specific for a particular haplotype in this interval and present in the extra DNA. On the other hand the size variation between haplotypes could be due to a difference in the number or the organization of the DRB genes. The haplotype-specific differences could have some bearing on the ability of chromosomes with different HLA haplotypes to recombine within the DQ to DR region.

As for class I, physical maps have been established for the class II region by using PFGE and these maps have been refined by analysis of YAC clones spanning the complete region (30, and see *Figure 6*). The most centromeric gene, KE3, lies ~ 150 kg from the DB cluster. The distance between the HLA-DP genes and the DNA gene is 80 kb. This gene and DOB are separated by 170 kb. In this interval a number of new genes are located such as the related class II genes HLA-DMA and -DMB, or the TAP and LMP genes involved in antigen processing through the class I pathway (32–35). Interestingly these new transcripts are associated with HTF islands mapped previously by PFGE analysis. The distance between the DOB to DQBA1 genes is approximately 180 kb with the DQB2, DQA2, DQB3, and DQB1 genes localized in this interval. As already mentioned, the distance between HLA-DQ and HLA-DRA is subject to variation from haplotype to haplotype. Nevertheless the average size of the interval can be estimated to be generally between 200–300 kg and even 400 kb in the DR8 haplotype (30).

6.3 The class III region

The physical organization of the class III region of the MHC by PFGE has established its size: 1100 kb (36–38, and see *Figure 5*). This region is very dense in genes encoding proteins unrelated to the classical class I or class II genes, although some are involved in the immune system, such as the complement components. As for the class I and II regions a number of diseases are associated with loci in the class III region. Several haplotypes have been studied by PFGE analysis of the class III genes (39). The DNA content seems not to vary except at the C4 and CYP21 loci. In fact the differences at these loci are due to the number and the length of the C4 genes (40) as established by RFLPs studies. Elsewhere, from the DRA gene (class II) to the

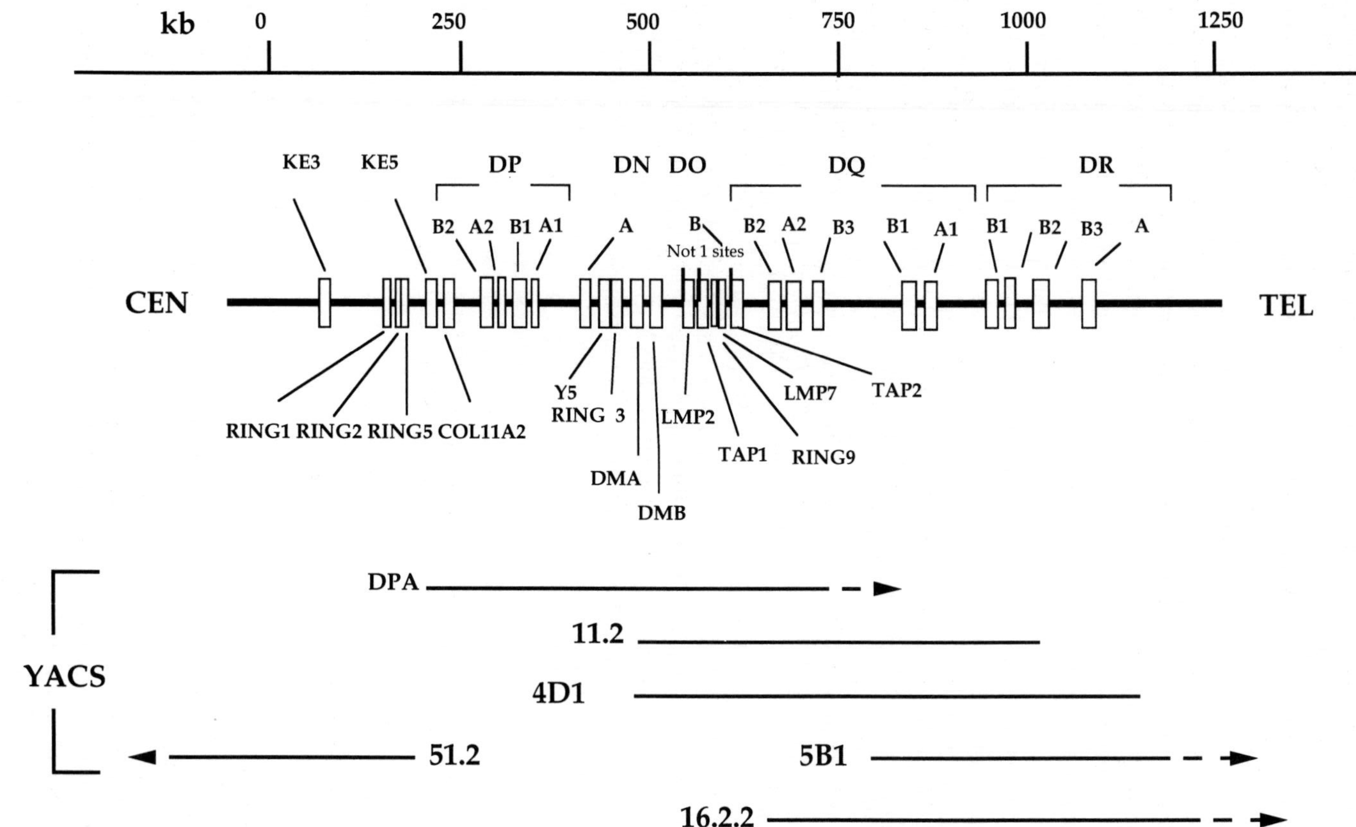

Figure 6. Molecular map of the class II region of the MHC with the derived YAC clones shown below.

HLA-B gene (class I) the DNA content is very consistent between different haplotypes. Tokunaga *et al.* (15) have indicated possible deletions of DNA within the class III region and telomeric to the TNF genes but these results were not confirmed.

A large number of HTF islands (38) were detected in the class III region by PFGE mapping. Genomic fragments from these islands were used to detect new transcripts. To date up to 40 genes have been located in the class III region to make this part of the MHC the most densely packed segment of the human genome.

7. Conclusion

The introduction of the PFGE technology in the second half of the 1980s was an important improvement for the construction of long-range physical maps. This technique provides a link between the genetic data in centimorgans and the conventional gel electrophoresis. It appeared possible to complete the map of a large region such as the MHC where probes were available. Furthermore at about the same time a new type of vector was developed: the YAC cloning system. These vectors contain large inserts which can be analysed by PFGE. By using these new technologies it was possible to establish a physical map of the MHC (even with comparison between different haplotypes) and to subsequently clone it in YACs. Complementary analysis of genomic and YAC DNA of a particular region by *in situ* hybridization for example (41), allows precise mapping. This information has been extensively used to search for new genes.

References

1 Schwartz, D. C. and Cantor, C. R. (1984). *Cell*, **37**, 67.
2. Carle, G. F. and Olson, M. V. (1984). *Nucleic Acids Res.*, **12**, 5647.
3. Carle, G. F., Frank, M., and Olson, M. V. (1986). *Science*, **232**, 65.
4. Southern, E. M., Anand, R., Brown, W. R. A., and Fletcher, D. S. (1987). *Nucleic Acids Res.*, **15**, 5925.
5. Serwer, P. (1987). *Electrophoresis*, **8**, 301.
6. Chu, G., Vollrath, D., and Davis, R. W. (1986). *Science*, **234**, 1582.
7. Hardy, D. A., Bell, J. I., Long, E. O., Liindsten, T., and McDevitt, H. O. (1986). *Nature*, **323**, 453.
8. Dunham, I., Sargent, C. A., Trowsdale, J., and Campbell, R. D. (1987). *Proc. Natl. Acad. Sci. USA*, **84**, 7237.
9. Caroll, M. C., Katzman, P., Alicot, E. M., Koller, B. H., Geraghty, D., Orr, H. T., *et al.* (1987). *Proc. Natl. Acad. Sci. USA*, **84**, 8535.
10. Ragoussis, J., Bloemer, K., Pohla, H., Messer, G., Weiss, E. H., and Ziegler, A. (1989). *Genomics*, **4**, 301.
11. Bird, A. P. (1987). *Trends Genet.*, **3**, 342.

12. Orbach, M. J., Vollrath, D., Davis, R. W., and Yanofsky, C. (1988). *Mol. Cell. Biol.*, **8**, 1469.
13. Maniatis, T., Fritsch, E. F., and Sambrook, J. (ed.) (1982). *Molecular cloning: a laboratory manual.* Cold Spring Harbor Laboratory Press, NY.
14. Campbell, R. D. and Trowsdale, J. (1993). *Immunol. Today*, **14**, 349.
15. Tokunaga, K., Saveracker, G., Kay, P. H., Christiansen, F. T., Anand, R., and Dawkins, R. L. (1990). *J. Exp. Med.*, **171**, 2101.
16. Shukla, H., Gillepsie, G. A., Srivastava, R., Collins, F., and Chorney, M. J. (1991). *Genomics*, **10**, 905.
17. Kahloun, A. E., Vernet, C., Jouanolle, A. M., Boretto, J., Mauvieux, V., Le Gall, J. Y., *et al.* (1992). *Immunogenetics,* **35**, 183.
18. Chimini, G., Boretto, J., Marguet, D., Lanau, F., Lanquin, G., and Pontarotti, P. (1990). *Immunogenetics*, **32**, 419.
19. Pontarotti, P., Chimini, G., Nguyen, C., Boretto, J., and Jordan, B. R. (1988). *Nucleic Acids Res.*, **16**, 6767.
20. Gruen, J. R., Goei, V. L., Summers, K. M., Capossela, A., Powell, L., Halliday, J., *et al.* (1992). *Genomics*, **14**, 232.
21. Schmidt, C. M. and Orr, H. T. (1991). *Hum. Immunol.*, **31**, 180.
22. Bronson, S., Pei, J., Taillon-Miller, P., Chorney, M. J., Geraghty, D. E., and Chaplin, D. D. (1991). *Proc. Natl. Acad. Sci. USA*, **88**, 1676.
23. Geraghty, D. E., Pei, J., Lipsky, B., Hansen, J. A., Taillon-Miller, P., Bronson, S., *et al.* (1992). *Proc. Natl. Acad. Sci. USA*, **89**, 2669.
24. Dunham, I., Sargent, C. A., Dawkins, R. L., and Campbell, R. D. (1989). *Genomics,* **5**, 787.
25. Kendall, E., Todd, J. A., and Campbell, R. D. (1991). *Immunogenetics*, **34**, 349.
26. Tokunaga, K., Kay, P. H., Christiansen, F. T., Saueracker, G., and Dawkins, R. L. (1989). *Hum. Immunol.*, **26**, 99.
27. So, A. K. L., Trowsdale, J., Bodmer, J. G., and Bodmer, W. F. (1984). In *Histocompatibility testing* (ed. E. Albert), pp. 565–8. Springer–Verlag, New York.
28. Todd, J. A. (1990). *Immunol. Today*, **11**, 122.
29. Inoko, H., Tsuji, K., Groves, V., and Trowsdale, J. (1989). In *Immunobiology of HLA* (ed. B. Dupont), Vol. II, pp. 83–6. Springer–Verlag, New York.
30. Ragoussis, J., Monaco, A., Mockridge, I., Kendall, E., Campbell, R. D., and Trowsdale, J. (1991). *Proc. Natl. Acad. Sci. USA*, **88**, 3753.
31. Ragoussis, J., Sanseau, P., and Trowsdale, J. (1991). In *Immunobiology of HLA* (ed. Sazazuki *et al.*), Vol. I, pp. 132–5. Oxford University Press.
32. Trowsdale, J., Hanson, I., Mockridge, I., Beck, S., Townsend, A., and Kelly, A. (1990). *Nature*, **348**, 741.
33. Glynne, R., Powis, S. H., Beck, S., Kelly, A., Kerr, L. A., and Trowsdale, J. (1991). *Nature*, **353**, 357.
34. Kelly, A., Powis, S. H., Glynne, R., Radley, E., Beck, S., and Trowsdale, J. (1991). *Nature*, **353**, 667.
35. Powis, S. H., Mockridge, I., Kelly, A., Kerr, L. A., Glynne, R., Gileadi, U., *et al.* (1992). *Proc. Natl. Acad. Sci. USA*, **89**, 1463.
36. Dunham, I., Sargent, C. A., Kendall, E., and Campbell, R. D. (1990). *Immunogenetics*, **32**, 175.
37. Dunham, I., Sargent, C. A., Dawkins, R. L., and Campbell, R. D. (1989). *J. Exp. Med.*, **169**, 1803.

38. Sargent, C. A., Dunham, I., and Campbell, R. D. (1989). *EMBO J.*, **8**, 2305.
39. Spies, T., Bresnahan, M., and Strominger, J. L. (1989). *Proc. Natl. Acad. Sci. USA*, **86**, 8955.
40. Kendall, E., Sargent, C. A., and Campbell, R. D. (1990). *Nucleic Acids Res.*, **18**, 7251.
41. Senger, G., Ragoussis, J., Trowsdale, J., and Sheer, D. (1993). *Cytogenet. Cell Genet.*, **63**, 49.

6

Derivation and analysis of MHC transgenic mice

ANNE-MARIT SPONAAS and ANDREW MELLOR

1. Introduction

Transgenesis is the term given to the process whereby cloned DNA is introduced into the germline of a host organism so that it becomes stably integrated into host chromosomes and is inherited from one generation to another. Procedures making transgenesis possible in mice were worked out in the early 1980s. Immunologists have been particularly active in exploiting the potential of this new genetic approach for studying immunological phenomena. There are two major reasons why transgenesis has been exploited for immunological research:

(a) The level and pattern of transgene expression in different tissues or during development can be studied and regulated with precision *in vivo*.

(b) Changes in phenotype due to transgene expression can be engineered exquisitely and consequent effects on complex biological phenomena can be monitored closely and assigned unequivocally.

In a sense, transgenic (Tg) mice represent a logical extension of immunogenetic research which revolutionized understanding of MHC genetics in the 1970s since Tg mice could be regarded as the ultimate congenic mouse with a single new (trans)gene expressed in the context of a genetic background from a well-characterized strain of laboratory mice. The role played by the MHC in immunological phenomena was barely understood, even in terms of cell–cell interactions, until immunogenetics provided the means to separate the genetic complexity of the MHC from the immunological role of MHC gene products. Even now, the role of the MHC in:

- development and selection of thymocytes
- antigen (peptide) presentation during immune responses to pathogens
- induction of tolerance to self-antigens
- autoimmunity
- rejection of tissue transplants

is still subject to intense scrutiny with Tg mice playing a pivotal role in attempts to understand the cellular and molecular events which lead to these complex phenomena. All these fields of enquiry have benefited from experiments involving MHC Tg mice carrying murine and human MHC class I and class II genes or recombinant DNA constructs derived from them. Because many immunological phenomena involve intimate and highly specific associations between T cell receptor (TCR) and MHC molecules, TCR–Tg mice, which express a single TCR clonotype on nearly all thymocytes and T cells, used in conjunction with appropriate MHC Tg mice, are extremely useful for immunological studies. In this chapter we intend to provide a practical account of how Tg mice are made (Section 2) and follow this with specific examples from our own research of how to test for expression of MHC (class I) transgenes (Section 3) and assess immunological status (Section 4) in a series of Tg mice. All Tg mice were made using inbred mice of the CBA/Ca strain ($H2^k$ haplotype) and all transgenes that we will refer to contain either human HLA-G or murine $H2\text{-}K^b$ structural genes. Of necessity, this account will be biased to our own research and experiences with Tg mice. MHC Tg mice expressing many types of murine and human (and other) MHC class I and II transgenes have been made in other laboratories as any search of a literature database will reveal. Interested readers should consult this literature for a broader view of the role and utility of MHC Tg mice in immunological research.

2. Production of transgenic mice

Transgenesis is achieved in two ways:

(a) Microinjection of purified linearized DNA into pronuclei of fertilized oocytes.
(b) DNA transfection of totipotent embryonic stem (ES) cells followed by transfer of transfected cells into blastocysts.

Manipulated oocytes (or blastocysts) are transferred into the oviducts (or uterus) of suitable (pseudopregnant) foster mothers by microsurgery. Detailed reviews and protocols of how to make, identify, and analyse transgene expression in Tg mice are available (1, 2). Both techniques are labour-intensive and require expertise in recombinant DNA technology, animal husbandry, micromanipulation, and microsurgery. Microinjection is the procedure of choice in most cases, partly for historical reasons, but also because ES cells must be carefully maintained so that they retain totipotency and because they are, at present, available from only one strain of mice (129/J) which limits choice of genetic background. This can be a very important consideration when making MHC Tg mice for use in immunological studies. Nevertheless, the ES cell route is now widely used to create so-called gene 'knock-out' mice. In this procedure, genes of interest are targeted using appropriate DNA constructs such that they are inactivated by insertional mutagenesis via homologous recombination in ES cells. Detailed discussion of this procedure is beyond the scope of this chapter but MHC I and II knock-out mice created in this way have proved to be extremely useful across many fields of immunological interest. Details about the microinjection procedure are given below. More extensive details are available for those wishing to set-up a facility from scratch (1, 2). What follows is a general discussion of our experiences of producing and analysing various MHC (and other) Tg mice made by microinjection at our facilities at the NIMR, Mill Hill, London since 1987. We have included many technical tips based on our direct experience and on our experiences of training others to use this procedure. An outline of the procedure we use is provided in *Figure 1*.

2.1 Facilities and equipment

A well-managed facility for animal production, breeding, and storage is essential for producing Tg mice. A facility for freezing and storing morulae is also desirable when large numbers and/or multiple lines of Tg mice are to be made for long-term studies. When planning microinjection sessions supplies of:

- young female mice (three to six week-old, depending on strain) for oocyte production
- stud males (six to eight week-old; use for up to three months)
- vasectomized male studs (six to eight week-old; use for up to three months)
- pseudopregnant females mated with them (six to eight week-old)

are required and must be carefully co-ordinated (*Figure 1*). A number of pieces of equipment are required for microinjection of DNA (*Table 1*) in addition to freezers and a CO_2-controlled environmental incubator for oocyte culture. Ideally, the microinjection facility should be located in an isolated (barriered) animal facility dedicated to the production of Tg mice to minimize the risk of infections in the mouse colony.

2.2 Oocyte production and manipulation

Superovulation (*Figure 1A*) is necessary to optimize oocyte yield and to synchronize ovulation and subsequent development of oocytes. Superovulation is achieved by injecting mice intraperitoneally with a preparation from pregnant mare serum (PMS), which contains follicle stimulating hormone, followed by human chorionic gonadotrophin (HCG) which induces ovulation. Dosage and timing of hormone injection should always be determined empirically since the age of females, the mouse strain, and the location of the colony all affect superovulation. For CBA/Ca mice at the NIMR, three to four week-old mice are best and hormones are administered between 1500–1800 hours on day 1 (PMS) followed by HCG between 1000 and 1200 hours on day 3. After administering HCG, females are mated overnight with stud males (*Figure 1B*)

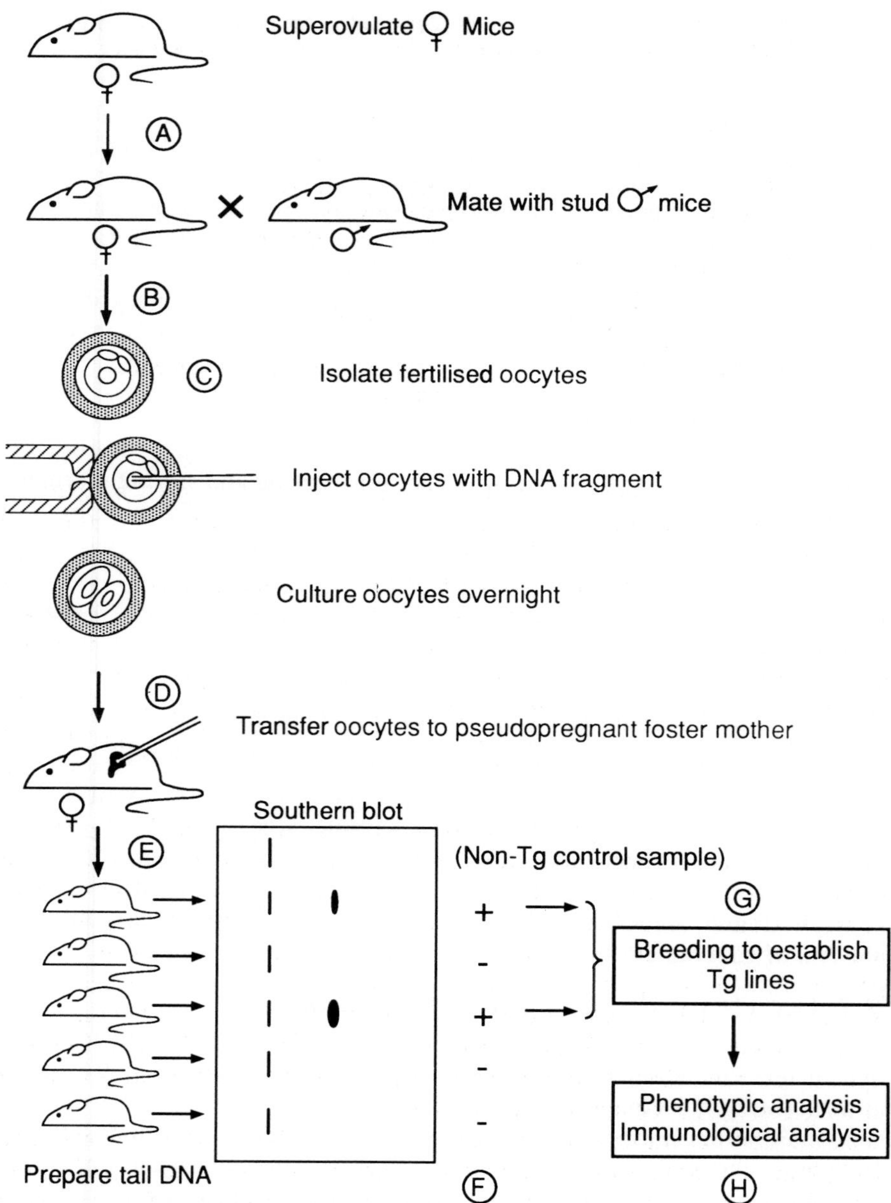

Figure 1. Procedure for producing Tg mice (details in text).

Table 1. List of equipment required for microinjection. For suppliers see reference 1 and 2

Equipment	**Required for:**
Binocular dissecting microscope	Oocyte isolation Transfers to oviducts
Microforge	Holding pipettes
Pipette puller	Microinjection needles
Two micromanipulators and micropipette holders	Manipulating and injecting oocytes
Inverted microscope with Nomarski optics	Pronuclear microinjection
Baseplate (custom built)	Microscope and micromanipulators
Air-cushioned table	Eliminating vibration

and inspected for vaginal plugs the next day (day 4). Fertilized oocytes are isolated from the oviducts of plugged females and are placed in culture in appropriate medium (*Figure 1C*) prior to microinjection with DNA. After injection, oocytes are transferred back to the oviducts of pseudopregnant foster mothers by microsurgery (sliding laparotomy). This can be carried out immediately after microinjection or can be carried out the next day after culture overnight when oocytes are at the two cell stage following the first division (*Figure 1D*). Litters are born 19 days after oocyte transfer (*Figure 1E*) and mice which have integrated copies of injected DNA in their chromosomal DNA (Tg founders, TGFs) are identified by genotypic analysis (*Figure 1F* and Section 2.4).

A summary of our experiences using CBA/Ca mice to make Tg mice is provided in *Table 2* which indicates average yield of oocytes and efficiency at each stage in the transgenesis procedure. Between four and ten TGF mice can be made from about 300 microinjected oocytes under optimal conditions. This number of oocytes can be isolated, microinjected, and transferred to pseudopregnant foster mothers on one day by an experienced person, although we also transfer oocytes the next day by which time they have divided once. The overall efficiency of transgenesis is affected by a number of factors (*Table 2*). Skill at micromanipulation and microsurgery must be acquired in order to attain a reasonable chance of success in transgenesis.

The conditions under which mice are kept affect breeding, superovulation, and plugging frequency by stud males. Apart from general quality of animal care, the timing of light cycles in the animal facility must be carefully regulated to obtain efficient superovulation and optimal plugging rates. Stud males should be caged separately and mated with females prior to their use for oocyte production since a high proportion of oocytes from females mated with 'virgin' males are not fertilized, which is presumably due to low sperm count in initial semen ejaculates. Plugging rates can be maintained at close to 100% if

Table 2. Production of transgenic mice (CBA/Ca strain)

Step	**Stage in procedure**	**Yield (optimal)**	**Factors**
A.	Superovulated females	20	Age, strain, light cycle, timing of injections
B.	Plugged females	20 (most)	Stud males, number of females per stud
Oocytes		300	(above 15/female or 7/oviduct)
C.	Fertilized	270 (90%)	Stud males, females
D.	Survive manipulation	240 (90%)	Media, quality of injection needles
E.	Pups born	36–48 (15–20%)	Females, number of oocytes per transfer (15–25)
Transgenic			
F.	Genotype	4–10 (10–20%)	Concentration of injected DNA
G.	Transmit	3–9 (90%)	Mosaicism, fail to breed
H.	Phenotype	3–9 (most)	Position effects, transgene stability

stud males are not used too frequently (two to three times a week, maximum), only one female at a time is placed with each stud male and males are replaced if they fail to plug females more than twice in succession. The plugging performance of each male should be recorded.

Another potential source of technical problems is the culture medium used to support oocyte development *in vitro* (M2 and M16; see refs 1 and 2 for details). Medium is made up batchwide from stock solutions and used within two weeks. Concentrated stocks of solutions used for medium preparation are stored as aliquots at −20 °C which should be discarded after one use. On one occasion, oocytes appeared fine before and after microinjection but pregnancies were not obtained. This problem was corrected by making medium from new stock solutions. The source of this problem was not entirely clear but the type of bovine serum albumin used and the method of storage of medium stocks were potential candidates. We have devised a simple protocol to avoid problems with culture medium. Prior to use in microinjection experiments, new batches of medium are tested for their ability to support oocyte development *in vitro* for a period of five days during which fertilized oocytes develop into blastocysts and even emerge (hatch) through the zona pellucida. Medium batches are only used if > 90% of oocytes proceed to this stage *in vitro*; we have had no problems with medium pre-screened in this way. One interesting corollary of this test is that most oocytes from inbred CBA/Ca mice will develop as far as the blastocyst stage *in vitro* despite the widely held view that oocytes from inbred strains cannot develop beyond the two-cell stage *in vitro*. We do not know how many other inbred strains are not subject to this two-cell block.

Most oocytes (> 90%) should survive microinjection unless glass injection needles are of poor quality. 'Good' injection needles are those that penetrate oocytes easily, do not block, do not burst oocytes, and that can be used to inject a large number of oocytes before such problems arise. Needles should be pulled on the day of microinjection as the glass creeps after pulling, although needles have been used successfully up to two weeks after being made. Even if all these problems are minimized, significant reductions in the expected number of pups born can occur. Sometimes foster mothers do not become pregnant even though vaginal plugs are evident after mating to vasectomized males and oocyte transfers proceed without obvious problems.

One of the most important factors affecting the overall efficiency of transgenesis is the concentration of DNA in the sample to be injected. This should be in the range of 1–2 μg/ml, although we use slightly higher concentrations when DNA fragments are very large or when several different DNA fragments are to be co-injected. In our experience, many people have difficulties with obtaining Tg mice simply because the concentration of DNA in the sample to be injected is too low. Frequently, this arises because concentrations based on spectrophotometer readings are inaccurate. For this reason, we always check the concentration of DNA samples to be microinjected by running them on ethidium bromide/agarose gels alongside control DNA samples which have been used previously to make Tg mice. If DNA concentration is optimal, the proportion of TGF mice obtained can be much higher than the average figure quoted in *Table 2*. For example, 50% of mice born carried a transgene in one experiment compared to the average of 10–20%. However, figures this high are not obtained often and other factors such as DNA purity, absence of toxic substances, or even toxic effects due to the transgene itself may affect frequency of transgenesis.

2.3 DNA for microinjection

DNA must be linearized by restriction enzyme digestion before microinjection since linear DNA is integrated into mouse chromosomes more efficiently than circular DNA. DNA fragments should be prepared free of all bacterial plasmid sequences since these can have detrimental effects on transgenesis and on transgene expression. Consequently, DNA must be purified (e.g. by agarose gel electrophoresis) prior to microinjection. We use agarose with a low melting temperature so that DNA can be extracted from gel slices by melting and phenol extraction. Toxic contaminants and particulate material are removed by passing the DNA over an appropriate resin column. We use 'Elutip-d' mini-columns (Cat. No. NA 010/0 supplied by Schleicher and Schuell) which are easy to use although DNA recoveries vary in the range 20–50%. DNA fragments are precipitated and dissolved in 5 mM Tris, 0.1 mM EDTA.

Dilutions of this stock solution are made (using the same buffer) and run on gels to determine DNA concentrations appropriate for microinjection (Section 2.2). Great care must be taken to avoid particulate material (e.g. dust particles from gloves) whilst preparing DNA as injection needles are extremely prone to blockage.

2.4 Identification of transgenic founders

Tg mice are identified by genotypic analysis of DNA prepared from tail biopsy samples (*Protocol 1*).

Protocol 1. Preparation of DNA from tails

1. Cut 3–4 mm of tail tip from 9–13 day-old mice into 1.5 ml microcentrifuge tubes. Add 500 μl of 'tail mix' (50 mM Tris–HCl pH 8, 100 mM EDTA, 100 mM NaCl, 1% SDS) and 10 μl of 20 mg/ml proteinase K and incubate at 55 °C overnight (can be done in 4–5 h).
2. Spin for 30 min at 13000 *g* in a (refrigerated) microcentrifuge. Decant supernatant into a fresh 1.5 ml tube.
3. Add 300 μl of isopropanol. Mix well. (Note: for a set of tubes do this step individually as DNA should not be left long in isopropanol.)
4. Fish out precipitated DNA immediately with hooked micropipette. Transfer to 200 μl 70% ethanol (EtOH) (leave in 70% EtOH until all samples in a set are precipitated).
5. Remove micropipette and dry briefly in air. Resuspend DNA in 100 μl TE (10 mM Tris pH 7.5, 1 mM EDTA) and leave to dissolve for a few hours at 55 °C or overnight at 4 °C.
6. Yield is usually about 100 μg. Use 20–25 μl for restriction enzyme digest for Southern blot. Digest in 100 μl volume. Use 40 μl digested DNA for gel electrophoresis. For PCR use 1 μl of stock DNA.

The presence of transgene sequences in DNA samples is determined using either Southern blot/hybridization analysis (*Figure 1F*) or PCR-based techniques. DNA prepared using *Protocol 1* can be subjected to either technique, although some restriction enzymes are sensitive to dirt and/or impurities in these rather crude DNA preparations. We always use enzymes (and suppliers) which are very reliable to avoid this potentially time-consuming problem; for example, *Bam*HI and *Bgl*II (from BCL) always work in our experience. It is possible to avoid the necessity for digesting DNA with restriction enzymes by filtering samples directly on to a nitrocellulose membrane ready for immediate hybridization to radiolabelled probes (dot or slot blots) as long as the probe does not cross-hybridize to endogenous murine DNA sequences. However, murine MHC transgenes (particularly class I genes) are not easy to detect by this simpler method unless unique (i.e. single copy) flanking DNA sequences are used as probes, or non-murine DNA sequences are incorporated into the transgene as a result of recombination since MHC genes belong to families of closely related sequences. Cross-hybridizing DNA sequences rarely pose a problem when digested DNA samples are analysed since transgene sequences will, in general, migrate differently on agarose gels used to fractionate digested DNA; this can be an advantage when trying to assess copy number or transgene homo- or heterozygosity since cross-hybridizing sequences provide a useful internal control for hybridization intensity. Although PCR analysis does not provide information about the relative copy number of transgene sequences in different DNA samples it is far less time-consuming than Southern blot/hybridization analysis. Consequently, we use PCR for routine identification of Tg mice but resort to Southern blots when we need to determine relative transgene copy number between different lineages carrying the same transgene, or when we need to identify 'homozygous' Tg mice which have transgenes on both chromosomes as a result of interbreeding (see Section 2.5). Successful PCR analysis depends largely on the design of the primers used. We use PCR primers which are 20 base pairs in length, have matched C + G contents in the range 55–65% (i.e. 11–13 C + G bases) with bias (if any) towards concentrating C + G bases at the 3′ end of the primer to ensure efficient priming of polymerization (computer programs are available for primer design). PCR products are designed to be in the range of about 200–1000 bp although there are some reports that more reliable and consistent results are obtained if slightly smaller PCR products (100–200) are generated. It may be useful to incorporate introns between the primer annealing sites so that RNA (as cDNA) and genomic DNA can be discriminated on the basis of PCR product size. This allows transgene detection and analysis of transgene transcription to be carried out with the same pair of primers. We recommend synthesizing primers for PCR detection of MHC class I transgenes which are homologous to highly polymorphic regions in exons 2 and 3 so that the second (~ 180 bp) intron lies between annealing sites. Again, close sequence relationships between some MHC (particularly class I) genes can complicate detection of MHC transgenes by PCR analysis and non-transgenic control samples should be tested for failure to provide templates for PCR primers used for MHC transgene detection.

2.5 Breeding transgenic lines

Tg mice generated as a result of oocyte injection are used to found Tg lineages by breeding to non-Tg mice of an appropriate strain (*Figure 1G*). We routinely breed our mice to non-Tg CBA/Ca partners as our projects are designed so that all Tg mice are made and maintained on the CBA/Ca background. Male founders are better than females since males can be placed with several CBA/Ca females to speed up the breeding process. Offspring of these matings are then typed and used for further breeding or actual experiments. It is advisable to select more than one Tg founder mouse per DNA construct because transmission of the transgene from founder to offspring is not guaranteed and transgene expression may vary due to differences in copy number and integration site position in individual founder mice. Provided that all germ cells in founder mice contain integrated transgenes, half of their offspring should be Tg mice. However, partial transgene mosaicism in germ cells often occurs leading to a reduced frequency of transgene transmission in the first generation. In some cases, no offspring inherit the transgene from the founder mouse, presumably because few germ cells, if any, contain stably integrated transgenes. Occasionally, transgene inheritance does not stabilize for the first, or even the second generation; this possibility must be considered before selecting mice of only one lineage for experimental use. Instability of transgene inheritance can also occur on relatively rare occasions when founder mice have two, or more, chromosomal sites where transgenes are integrated. Segregation then leads to genotypic and, potentially, to phenotypic differences between individuals in a given lineage. Alternatively, illegitimate recombination, leading to deletion of DNA, within clusters of linked transgene sequences at a single integration site can generate phenotypic instability from one generation to the next. When Tg mice are to be used in a long-term project it is advisable to breed them to homozygosity and to freeze morulae so that the line can be resurrected at a later date if existing colonies are lost on account of breeding failure or infection. After the first generation of Tg mice has been born we set-up new matings using male and two female Tg offspring of the founder mouse; we discontinue matings involving the founder unless s/he has passed on the transgene to a large proportion (i.e. close to 50%) of her/his offspring. We usually stop all matings in which non-Tg mice are involved to increase the proportion of homozygous mice in the second generation. 25% of mice born of two heterozygous Tg parents should be homozygous for the transgene and these are very useful for ongoing breeding since all their offspring will be Tg mice. Furthermore, if both parents are homozygous the lineage breeds true for the transgene and, at least in theory, typing can be discontinued. In practice, inherent uncertainties in identifying homozygous Tg mice often mean that typing is continued at least for another generation. Occasionally, genetic instability at the transgene locus when both chromosomes carry a transgene or phenotypic lethality due to increased dose of transgene products might occur. In our experience with mouse and human MHC class I transgenes lethal effects due to transgene expression are not manifested. Very rarely transgenes cause lethality as a consequence of integration into a genetic locus which provides an essential gene function. Frequently, genetic background is of critical importance in immunological research because of the influence that polymorphic gene products (e.g. major and minor histocompatibility genes and *mtv* (mammary tumour virus) related sequences encoding superantigens) have on the repertoire of T and B lymphocyte receptor specificities in individual mice. Hence genetic background should be carefully considered when making MHC Tg mice for use in long-term projects. For example, tissue transplantation and autoimmune conditions can be susceptible to multiple effects encoded at segregating genetic loci. Unfortunately, the majority of Tg mice have been (and are still being) made using second generation (F2) oocytes from two laboratory inbred mouse strains with different genetic backgrounds (1, 2). This means that each Tg founder mouse possesses a unique set of background genes inherited as segregating sets of genes derived from both parental strains and can be homozygous, hemizygous, or null for alleles from each parental strain. This may not present a great problem since heterogeneity at loci which exert the most prominent influence on T and B cell repertoire selection (i.e. MHC and *mtv* loci) can often be resolved in a single generation by judicious choice of mating partners and typing. However, Tg lineages derived from a single founder mouse and maintained by repeated backcrossing to partners from a particular inbred strain will not acquire true 'inbred' genetic backgrounds until the 12th and subsequent backcross generation. This can take up to three years to achieve and, where influences from background genes must be eliminated completely, this can present a major problem. Largely for this reason we chose to generate Tg mice using 'inbred' oocytes where both parents are from one inbred strain. This is feasible for several inbred strains including CBA/Ca (3) and C57BL strains (4) although they may be more difficult to work with than oocytes made from non-inbred mice. For example, the C57BL/10 (B10) strain at the NIMR is much harder to use for transgenesis than CBA/Ca mice because of lower oocyte yields, more erratic mating, and particularly because oocytes are more susceptible to bursting during culture and after injection. Susceptibility to manipulations *in vitro* is due, at least in part, to genes controlling resistance to hyaluronidase which is used to facilitate removal of cumular cells adhering to oocytes (5). One general indication of the potential of an inbred strain for transgenesis is the ease with which they breed as a colony since some, apparently identical, inbred strains breed better in some Institutions than others.

3. Assessing MHC transgene expression

After Tg mice have been identified by genotypic analysis some form of phenotypic analysis is normally carried out. For those interested in gene expression, Tg mice offer a unique system of investigation *in vivo* which complements *in vitro* techniques such as DNA-mediated gene transfection. Often, it is not necessary to establish long-term breeding colonies of Tg mice if studies on the regulation of gene expression are the exclusive research goal. Instead, Tg founder mice can be subjected to rigorous analyses using a variety of techniques to assess transcription patterns in various tissues of adult Tg mice or during development. Usually, the initial goal of such projects is to mimic the normal pattern of gene expression using a transgene consisting of cloned DNA incorporating the gene of interest. Once this goal has been realized, the next stage is to modify the original transgene (usually by progressive deletion of DNA sequences) until the pattern of transgene transcription is modified. In this way, DNA elements controlling gene expression are identified. Further characterization may involve making more Tg mice using transgenes in which potential control elements identified by deletional analysis are recombined with other genes to demonstrate their regulatory role. This may involve assays for protein expression and for altered immunological functions *in vivo* as well as *in vitro*. What follows is an account of our experiences with determining the pattern of transgene expression in Tg mice carrying either a human MHC class I transgene (HLA-G) or various recombinant transgenes containing a murine MHC class I gene (H2-K^b).

3.1 Transcription of MHC transgenes

RNA samples for transcriptional analysis are prepared from tissues extracted from freshly killed Tg mice. There are several ways to isolate RNA from tissues and all require care to avoid degradation of RNA by RNases from tissues or contaminated reagents used in the preparation. We prefer to use methods based on the use of guanidinium isothiocyanate which inactivates enzymes rapidly when cells are lysed. The procedure we use for isolating RNA from small tissues (e.g. embryos) is shown in *Protocol 2*.

Protocol 2. RNA preparation from tissues (small scale)

1. Mince tissue (up to 500 mg tissue; increase volumes for larger tissues) and homogenize (Polytron) in 5 ml GT mix on ice keeping sample as cold as possible. GT mix: 4 M guanidinium thiocyanate, 25 mM Na citrate pH 7, 0.5% Sarcosyl, 0.1 M 2-ME (prepare stock without 2-ME; store frozen in aliquots. Add 2-ME when required).
2. Add 0.5 ml 2 M NaOAc pH 4 and extract with 5 ml phenol:chloroform.[a]
3. Centrifuge 10 000 *g* for 20 min at 4 °C.
4. Remove aqueous phase and precipitate with 1 ml of isopropanol. Leave at −20 °C for at least 1 h. Spin 10 000 *g* for 20 min at 4 °C.
5. Dissolve RNA pellet in about 1.2 ml GT mix. Transfer to two 1.5 ml microcentrifuge tubes. Add 1 vol. isopropanol. Precipitate at −20 °C for 1 h.
6. Spin 10 min 4°C. Resuspend pellet in 70% EtOH. Dry. Take up in 0.5% SDS or DEPC treated water.

[a] Cleaner RNA can be prepared by placing lysate in GT mix over CsCl cushion gradients.

This procedure can be scaled up for larger tissues (e.g. spleen, thymus, liver, kidney). Once RNA samples have been prepared a number of methods can be used to detect the presence of specific transcripts. Northern blot analysis (protocol in ref. 6), is the least sensitive method available and for some immunological applications may be inappropriate. In addition, the close sequence relationship and similar size of MHC gene transcripts can make this method of detection difficult to use for detection of transcripts of MHC transgenes, although the use of synthetic radiolabelled oligonucleotides can circumvent the problem of specificity. A more sensitive technique is nuclease S1 protection analysis (6). This method is quite difficult to carry out and, again, may not be specific enough to avoid detection of transcripts from host MHC genes when attempting to detect murine MHC transgene transcription. The most sensitive method of detection is a PCR-based technique in which RNA is first converted to cDNA and then subjected to PCR amplification using appropriate pairs of primers (*Protocol 3*).

Protocol 3. Detection of transgene transcription by RT-PCR

A. *First strand cDNA synthesis*

1. Denature RNA (1 μg) in 7.3 μl DEPC treated water at 70°C, 1 min. Quench on ice.
2. Add 5 μl of 2 mM dNTPs (Pharmacia LKB), 4 μl 5 × H-RT buffer (BRL), 2 μl of 0.1 M DTT (BRL), 0.2 μg oligo(dT) primer (Pharmacia LKB), 1 U of RNasin (BCL), and 0.5 μl MMLV reverse transcriptase (BRL). Incubate at 37 °C 1 h.
3. Heat to 95 °C for 5 min. Add 80 μl DEPC treated water.
4. Use immediately for PCR or store at −70 °C.

B. *PCR (cycling conditions may vary depending on primer design)*

1. To 10 μl of first strand cDNA add 1 μl each of 12.5 μM sense and antisense primers and 3 μl of DEPC treated water.
2. Overlay with mineral oil. Heat to 95°C, 1.5 min in PCR machine to denature.
3. Add 15 μl containing 3 μl 10 × PCR buffer (15 mM $MgCl_2$, 100 mM Tris pH 9.0, 500 mM KCl, 1% Triton X-100), 0.6 μl of 10 mM dNTPs (Pharmacia LKB), water, and 0.5 μl (2.5 U) *Taq* polymerase (Perkin-Elmer Cetus).
4. Amplify by PCR (we use 30 cycles of 94 °C, 1 min; 55–65 °C, 2 min; 72 °C, 2 min; and end with 72 °C, 7 min).
5. Analyse PCR products (10–15 μl) by EtBr/agarose gel electrophoresis.

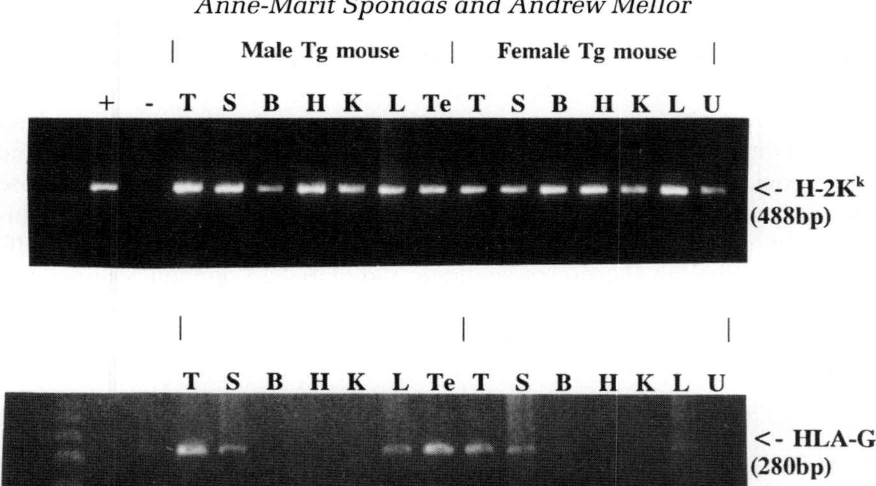

Figure 2. Transcripts of HLA-G transgene in tissues from adult Tg mice detected by RT–PCR. RT–PCR products from H2-K^k (control) templates (top) and HLA-G transgene (bottom) are indicated by arrows. For key to RNA sources see text. (Courtesy of Dr A. Horuzsko.)

An example of the use of the RT-PCR technique to detect transgene expression during embryonic development and in adult Tg mice carrying a human MHC class I transgene, HLA-G, is shown in *Figure 2*. In this example, RNA samples extracted from tissues of male and female Tg mice carrying an HLA-G transgene were tested for the presence of HLA-G transcripts. Transcripts of the HLA-G transgene were detected in RNA from spleen (S), thymus (T), and liver (L) of both mice, and in testis (Te) RNA from the male mouse, but were not detected in RNA from brain (B), heart (H), kidney (K) of either Tg mouse, or RNA from uterus (U) of the female mouse. In contrast, transcripts of the murine MHC class I gene H2-K^k were present in all samples indicating that the absence of HLA-G transcripts in some samples arises from tissue-specific regulated expression of the HLA-G transgene and is not due to degradation of RNA in those samples.

3.2 MHC protein expression

MHC molecules can be detected in two ways: using either monoclonal antibodies (Section 3.2.1) or T cells specific for MHC products (Section 3.2.2). What follows is an account of methods we have used to detect expression of murine H2-K^b molecules on the surface of cells from Tg mice called CBK (3) (H2), KAL (7) (α-lactalbumin), Kβ (3) (β-globin), or CD2K^b (8) (CD2). Transcriptional promoters present in each transgene are indicated in the brackets. These promoters are active in different cell types and, consequently the pattern (tissue-specificity) and/or timing (e.g. during embryonic development) of H2-K^b expression varies in each type of Tg mice.

3.2.1 Detection using monoclonal antibodies

Monoclonal antibodies are highly specific reagents for detecting particular epitopes. A range of techniques designed for this purpose can be used and include:

- immunoprecipitation
- Western blotting
- enzyme-linked immunoabsorbance assay (ELISA)
- immunohistological analysis
- cytofluorometric analysis

We have not used immunoprecipitation or Western blotting in our research work involving MHC Tg mice. However, these techniques are useful for studying the biosynthesis or biochemistry of protein molecules and are used extensively by immunologists interested in antigen presentation involving MHC molecules.

i. Immunohistological analysis and immunofluorescence

Immunohistological analysis of tissue sections obtained from MHC Tg mice is a very useful way of detecting expression of MHC transgenes. Cells stained with antibodies can be visualized with a number of reagents. For example, antibodies coupled to fluorescent dyes can be visualized by fluorescence microscopy. If staining for MHC transgene expression is carried out in the presence of other antibodies which stain particular cell types (such as thymocytes, T cells, or B cells) it is possible to verify whether or not transgene expression takes place in these cells. In the example shown in *Figure 3*, thymus sections from control (*Figure 3A*, B10, positive control; *Figure 3B*, CBA, negative control) and Tg (*Figure 3C*, CD2K^b) mice have been stained with two monoclonal antibodies, one specific for H2-K^b (green stain) and the other specific for keratin (red stain), a marker for thymic epithelial cells. As expected, no green staining cells are present in the CBA thymus, whereas co-expression of the H2-K^b and keratin on epithelial cells in B10 thymus generates yellow fluorescence (green staining thymocytes can also be seen in this section). In contrast, the staining pattern on thymus sections from the CD2K^b Tg mouse is different on account of an increase in the level of H2-K^b expression on thymocytes and the possible absence of H2-K^b expression on epithelial cells in these mice (note however, that it is difficult to detect red staining cells, probably due to overlap of thymocytes and epithelial cells; confocal microscopy would help resolve whether epithelial cells fail to express H2-K^b). The drawback of this technique is that it is relatively insensitive compared to other methods of protein detection. For example, we were unable to detect expression of H2-K^b molecules in the thymus of some Tg mice

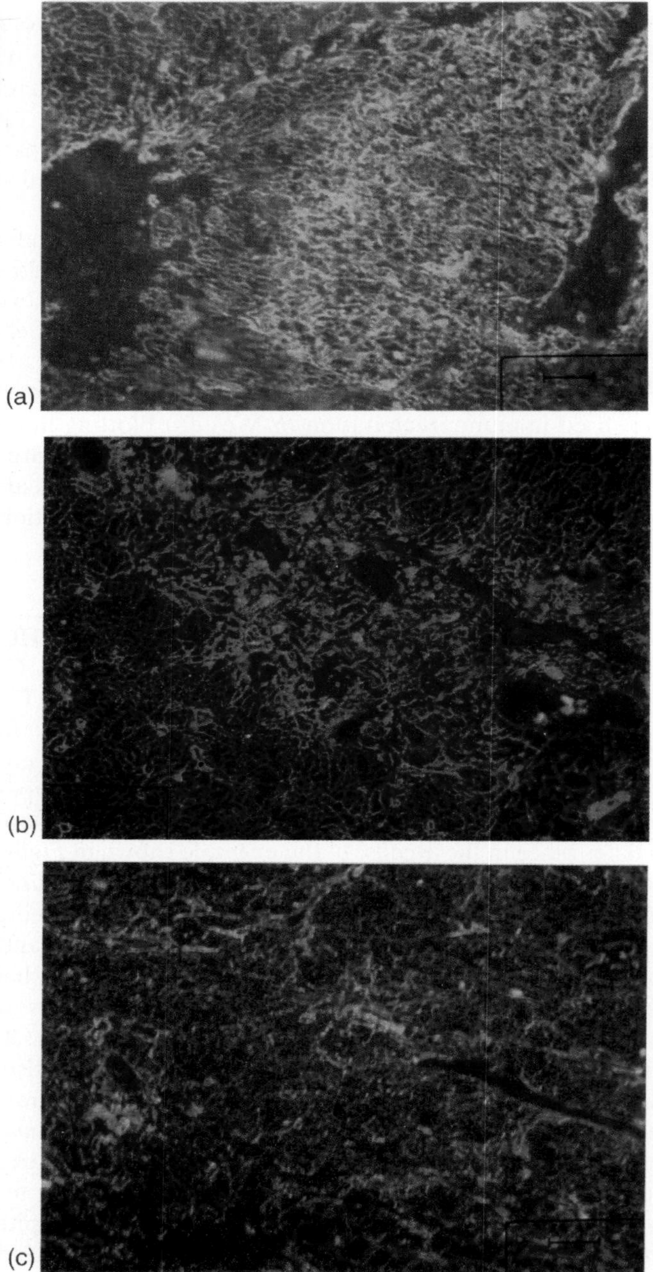

Figure 3. Immunoflourescent staining of thymus sections from (A) C57BL (B10), (B) CBA, and (C) Tg CD2K^b-3 mice. Sections were stained with an anti-keratin antibody coupled to PE (red) and an anti-H2-K^b antibody coupled by FITC (green).

despite the fact that these mice could be shown to have acquired T cell tolerance to H2-K^b in the thymus (3, 7). Nevertheless, immunofluorescence analysis is a very powerful method of detecting small subpopulations of cells which express MHC molecules as long as the level of expression on these cells is above the threshold of detection.

ii. *Flow cytofluorometry (FACS analysis)*

Cytofluorometric analysis of cell suspensions is a quick and reliable way of detecting protein expression on cells such as lymphocytes, although many other cell types are amenable to this type of analysis. In brief, cells in homogeneous suspensions are stained with monoclonal antibodies conjugated to fluorochromes (*Protocol 4*) and are passed through a laser beam which induces emission of light of precise wavelengths which is detected, digitized, and analysed electronically using a computer.

Protocol 4. Immunofluorescent staining of cells for cytofluorometric analysis

Reagents

- Monoclonal antibodies directly conjugated to fluorescent dyes (or to biotin in which case a second layer of streptavidin–fluorescent dye is required)
- Wash buffer: PBS with 2% FCS or 1% BSA, 0.02% sodium azide (to prevent capping)

Method

1. Prepare cell suspensions in wash buffer and place in 96-well microtitre plates on ice. Use between 10^5 and 10^6 cells or 100 μl (minimum) mouse blood in heparin when analysing PBLs. Spin at 1200 *g* 5 min. Invert plates to remove supernatants. Resuspend cells gently in residual buffer. RBL should be removed from mouse blood.
2. Add antibody(ies) or antibody conjugates in wash buffer (50 μl) to cells. Incubate on ice for 20 min.
3. Spin and wash. Resuspend cells as before.
4. Add second antibody (e.g. anti-mouse Ig–FITC) or streptavidin–PE if using biotin conjugates. Incubate for 15 min on ice.
5. Wash twice. Resuspend in wash buffer.
6. Analyse by cytofluorometry (e.g. on FACSCAN® analyser).

Machines for carrying out sophisticated cytofluorometric analyses are available in most well equipped immunology research laboratories. This technique allows:

(a) Definition of cell characteristics, such as size and granularity, as judged by light scattering properties (forward and side scatter, respectively).

(b) Accurate and rapid quantitation of fluorescence intensity for each cell analysed.

(c) Simultaneous definition of up to three or four fluorescent dyes defining different markers.

(d) Sorting of cells based on size, granularity, and fluorescence characteristics.

Data obtained from cytofluorometric analyses can be presented as:

(a) Histograms (one parameter) showing the number of cells (*y* axis) with a given relative fluorescence intensity (*x* axis, channel number).

(b) Dot plots (two parameters) in which each cell (event) is represented on a two-dimensional dot matrix plot.

(c) Contour plots in which lines representing dot densities replace dots in two-dimensional plots.

When cells are stained with three fluorescent dyes simultaneously it is common practice to present results in the form of a two-dimensional dot (or contour) plots which allow distinct populations to be electronically defined (or gated) so that fluorescence profiles of cells falling into each gated region can be separately analysed for third-colour fluorescence expressed in the form of a histogram, An example of the use of FACS analysis to detect expression of H2-K^b on peripheral blood lymphocytes (PBLs) of Tg and control mice is shown in *Figure 4*. The results are presented as contour plots on a two-dimensional display; the vertical axis represents red (PE, phycoerythrin) staining due to H2-K^b expression, and the horizontal axis represents green (FITC, fluorescein-isothiocyanate) staining due to expression of the T cell marker Thy-1. The events displayed were selected on the basis of forward and side scatter characteristics typical of lymphocytes. T cells appear on the right (Thy-1^+ and B cells on the left (Thy-1^-). All lymphocytes express H2-K^b in control B10 mice whereas no cells express H2-K^k in control CBA/Ca mice, as expected. T cells in both CD2K^b Tg mice analysed express H2-K^b whereas B cells express little, if any, H2-K^b. It can also be seen that the level of H2-K^b expression differs in the Tg mice (this is due to Tg copy number).

3.2.2 Detection using functional (T cell) responses

T cells recognize and respond to peptides presented as complexes with MHC molecules on the surface of presenting (or target) cells. Assays based on T cell responses are complementary to studies based on antibody detection and are of direct relevance for research into induction of T cell tolerance and autoimmunity. T cell responses can be measured *in vitro* following mixed lymphocyte culture (MLC) in which responder T cells obtained fresh from peripheral lymphoid organs of responder mice are placed in culture with inactivated stimulator cells (*Figure 5*, and *Protocol 5*). Responder cells can be

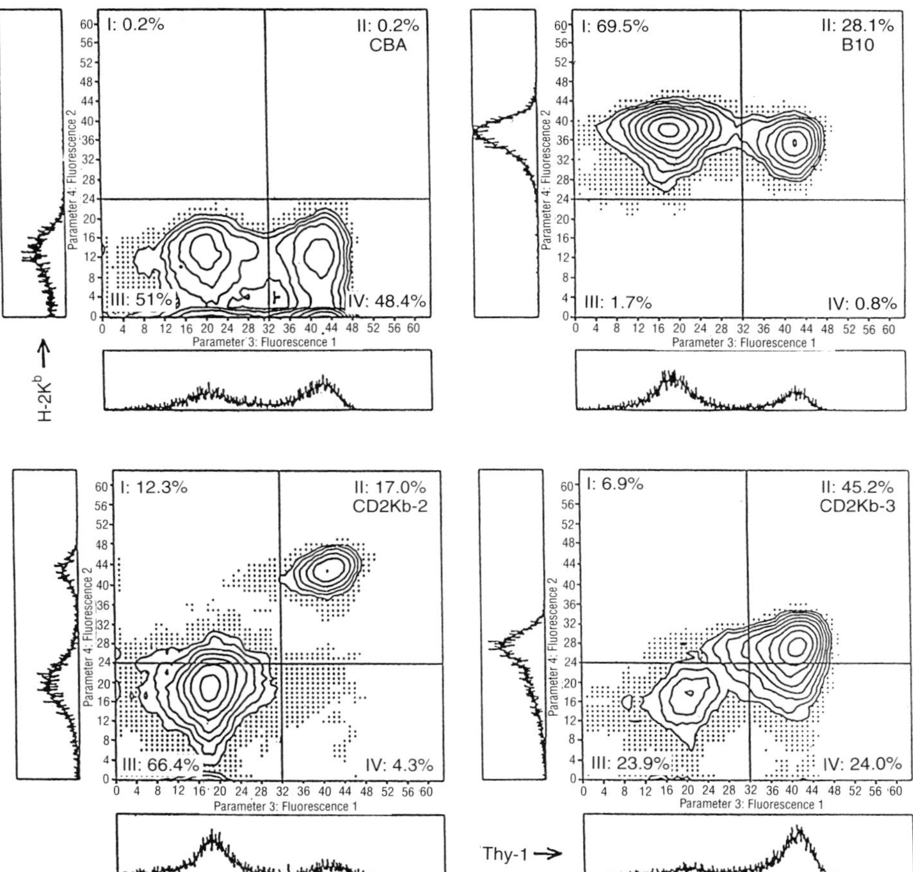

Figure 4. Cytoflourometric analysis of H2-K^b expression on PBLs. PBLs were stained with anti-Thy-1–FITC conjugate (horizontal axis) and anti-H2-K^b (B8.24.3) conjugated to biotin and developed with streptavidin–PE (vertical axis). Results are presented as contour plots with histogram profiles on each axis.

obtained from non-Tg, MHC Tg mice, or from TCR–Tg mice in which all T cells express the same TCR clonotype (idiotype). TCR–Tg mice are well suited to this type of analysis because they provide a source of T cells with precisely defined antigenic specificity, and consequently respond more vigorously in MLC than T cells obtained from non-TCR–Tg mice. We have generated several lines of TCR–Tg mice in which transgenic TCR molecules confer reactivity to H2-K^b (e.g. BM3 TCR–Tg mice) (9). We use T cells from these mice to assess H2-K^b expression and study the immunological status of H2-K^b Tg mice (Section 4.1).

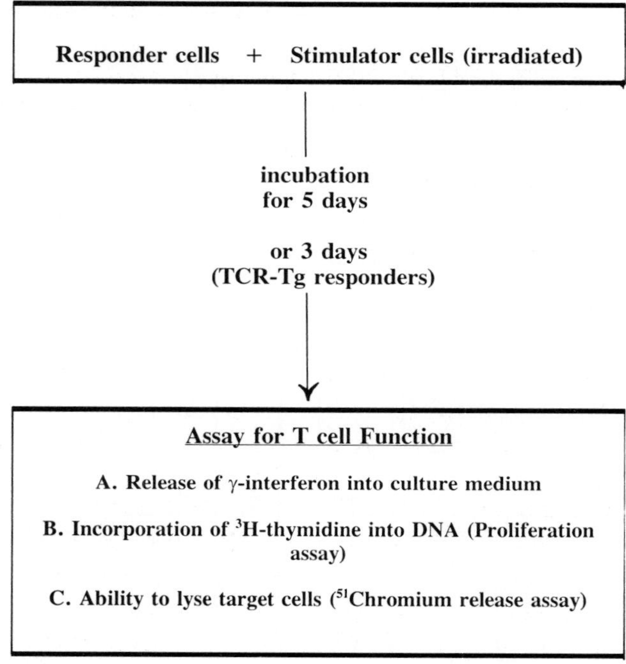

Figure 5. Standard procedure for MLC.

Protocol 5. Mixed lymphocyte cultures

Reagents

- Wash medium: air-buffered IMDM (Iscove's modified Dulbecco's medium) or BSS (balanced salt solution) and 2% FCS
- Culture medium: bicarbonate-buffered IMDM, 10% FCS, 10 mM glutamine, 100 U/ml penicillin, 100 μl/ml streptomycin, 5×10^{-4} M 2-mercaptoethanol

A. *Prepare responder cells (from spleen and/or lymph nodes)*

1. Dissect and tease spleen/lymph nodes to make homogeneous cell suspension. Wash cells twice in wash medium.
2. Count cells in 0.2% (w/v) Trypan blue (ignore dead, stained, cells) and resuspend at 4×10^6 cells/ml in culture medium.

B. *Prepare stimulator cells (from spleen and/or lymph nodes)*

1. Dissect, tease, and wash cells as above.
2. Remove red cells (from spleen) by water or ammonium chloride lysis.
   (a) Water lysis: add 4.5 ml sterile water to pelleted spleen cells. Add 0.5 ml 10 × BSS after 10 sec. Mix and add 15 ml wash medium. Spin and wash.
   (b) Ammonium chloride lysis: incubate spleen cells with 5 ml warm 0.83% NH_4Cl for 5 min at 37 °C then add 15 ml wash medium. Spin and wash.
3. Count cells. Resuspend at 4 × 10^6 cells/ml in culture medium.
4. Irradiate (2000 Rads) cells (or treat cells with mitomycin C if no radiation source is available).

C. *MLC set-up*

1. Mix 4 × 10^6 responder and 4 × 10^6 irradiated stimulator cells/well in 24-well tissue culture plates (Nunc, Falcon). Use 10^6 stimulators for TCR–Tg responders.
2. Incubate three to six days at 37 °C in a humidified CO_2 incubator. Optimal time depends on type of assay (see below) and should be tested empirically. For non-TCR–Tg mice five days is optimal (anti-MHC class I responses); for TCR–Tg responders three days should suffice.

After a period of three to six days, depending on the type of read-out assay to be used and the source of responder T cells, responses to antigen stimulation are assessed by measuring cytokine release (γ-IFN, IL-2, IL-4, IL-10, etc.) into culture supernatants (e.g. *Protocol 6*), proliferation (*Protocol 7*), or their ability to lyse target cells specifically (*Protocol 8*).

Protocol 6. Measuring release of γ-interferon by T cells

Reagents

- Two anti-murine γ-IFN antibodies; AN18 (10) and R4-6A2 (ATCC code HB 170) (one antibody should be biotinylated)
- Streptavidin–horseradish peroxidase (S-HRPO, Southern Biotechnology)
- Coating buffer: 0.2 M borate pH 8.5
- Blocking buffer: PBS with 5% FCS and 5% horse serum
- Wash solution: PBS with 0.05% Tween 20
- ABTS substrate (Sigma A-1888)
- 10 mM phosphate buffer pH 6.2

Method

1. Coat 96-well flat-bottomed flexiplates (Dynatech) with anti-γ-IFN antibody diluted to 10 μg/ml in coating buffer overnight at 4°C in a humidified box.

Protocol 6. *Continued*

2. Wash plates. Place blocking buffer in wells for 1 h at room temperature. Flick off buffer.
3. Place 200–100 μl supernatant from MLC in wells for at least 2 h at room temperature. Wash three times.
4. Place 50 μl of biotinylated second anti-γ-IFN antibody (5 μg/ml) in each well. Incubate 1 h at room temperature. Wash three times.
5. Add 50 μl of streptavidin–HRPO (diluted 1:500 in PBS) to each well. Incubate for 1 h at room temperature. Wash three times (use phosphate buffer for the final wash).
6. Dissolve 27.4 mg/ml of ABTS substrate in phosphate buffer. Add 25 μl/ml of 0.1% H_2O_2. Add 50 μl substrate mix to each well. Colour develops within 5–10 min.
7. Read optical density (414 nm) in an ELISA plate reader. Add substrate alone as a blank. Prepare a standard curve using recombinant γ-IFN to convert to concentrations.

Protocol 7. Assay for T cell proliferation

Reagents

- As for *Protocol 5*
- [^{3}H]Thymidine (Amersham)
- Fetal calf serum should be screened for low background in proliferation prior to use

Method

1. Set-up MLC (*Protocol 5*) but mix 5×10^4 responder and 5×10^5 stimulator cells/well in quadruplicate in a flat bottomed multiwell plate. Lymph node cells are best: if spleen cells are used, remove B cells before setting-up MLCs.
2. Incubate MLC for three to five days. It is important to determine which period gives optimal levels of proliferation.
3. Pulse with 1 μCi/well of [^{3}H]thymidine for 18 h.[a]
4. Harvest cells using an automated cell harvester.
5. Determine [^{3}H]thymidine uptake by scintillation counting and calculate mean and standard deviations.
6. Include syngeneic stimulator controls for determination of background levels of proliferation.

[a] When using T cells from TCR–Tg mice mix 2×10^4 responder and 4×10^5 stimulator cells/well in quadruplicate. Culture for three to four days and pulse with 1 μCi[^{3}H]thymidine for 12 h only before harvesting.

Protocol 8. Chromium release assay for measuring specific cytotoxic activity

Reagents

- As for *Protocol 5*
- Concanavalin A (Pharmacia LKB)
- LPS (Sigma)
- Assay medium: IMDM + 5% FCS.
- Sodium ^{51}chromate (Amersham)

A. *Preparation of target cells (spleen Concanavalin A blasts)*

1. Dissect spleens to prepare a homogeneous cell suspension. Wash cells twice in wash medium.
2. Place 10^6 cells/well and add 3 ml/well of culture medium in a 12-well tissue culture plate (Nunc, Falcon) with Con A at 1–2 μg/ml. Incubate three to six days at 37 °C in CO_2 incubator.
3. Target cells can be prepared from splenic B cells using LPS instead of Con A. Tumour cells and fibroblasts also make good target cells.

B. *Chromium release assay*

1. Harvest and wash target cells twice. Count and pellet target cells (not more than 5×10^6 cells/tube). Resuspend and add 100 μCi ^{51}Cr. Incubate for 90 min at 37 °C.
2. Harvest, wash (twice), and count responder cells. Resuspend at 3×10^6/ml in assay medium.
3. Add 150 μl of responder cells to each of three wells in a multiwell plate. Take 50 μl out into adjacent three wells and add 100 μl assay medium. Repeat this dilution twice.
4. Wash target cells and resuspend at 10^5 cells/ml in assay medium. Add 100 μl (10^4) to all wells containing diluted responders. This creates responder:target ratios of 30:1, 10:1, 3:1, and 1:1. Centrifuge (optional) multiwell plates for 1 min at 1000 *g*. Incubate for three to five h in CO_2 incubator; some targets lyse faster than others and optimal incubation times should always be determined.
5. Set-up controls for maximum release (add 100 μl 5% Triton X-100 to 100 μl target cells) and minimum (spontaneous) release (add 100 μl assay medium to 100 μl target cells). Incubate alongside experimental samples.
6. Harvest up to 100 μl supernatant (25 μl if using a Pharmacia beta plate 1205 counter) and measure ^{51}Cr released by scintillation counting.
7. Per cent specific chromium release is calculated as:

   [experimental − minimum]/[maximum − minimum] × 100%.

   Regression analysis can be performed as described by Simpson and Chandler (11).

Table 3. T cell proliferation assay

Responder	**Stimulator**	**[^{3}H]thymidine uptake ($\times$ 10−3)**
CBA	CBK	15.7
CBA	CBA	8.5
CBA	BALB/c	36.1
BM3	CBK	100.0
BM3	CBA	0.6

Tabld 4. Assay for T cell cytotoxicity[a]

		Target cells (% specific ^{51}Cr release)		
Responder	**Stimulator**	**CBK**	**BALB/c**	**CBA**
CBA	CBK	30	0	0
CBA	BALB/c	0	53	0
BALB/c	CBA	0	0	40
BM3	CBK	54	NT	0
BM3	BALC/c	45	NT	13

[a] Figures indicate per cent specific lysis at responder:stimulator = 10:1.

Examples of using T cells to assess expression of H2-K^b in Tg mice (CBK) are presented in *Table 3* (proliferation assay) and *Table 4* (^{51}Cr release assay). BALB/c stimulators are used as a 'third party' control. Note that H2-K^b-specific T cells from TCR–Tg mice (BM3) respond more vigorously than non-Tg CBA responders but there is some cross-reactivity since CBK targets are lysed by BM3 responders stimulated with BALB/c cells. T cells from BM3 TCR–Tg mice are very convenient for measuring expression of MHC molecules on various types of cells or tissues. We have tested the sensitivity of this method of detecting H2-K^b expressed on cells from CBK mice by measuring responses when T cells from BM3 mice are incubated with mixtures of CBA and CBK cells in various proportions (*Figure 6*). The threshold of detection in all assays was in the range 2.5–5%. However, younger mice produce higher levels of lytic activity than older mice (*Figure 6A*).

3.2.3 Detection of MHC expression by skin grafting

MHC molecules provoke tissue rejection between genetically mismatched individuals. This can be used as a very sensitive assay for MHC expression on tissues grafted from MHC Tg mice to non-Tg mice (*Protocol 9*) (12).

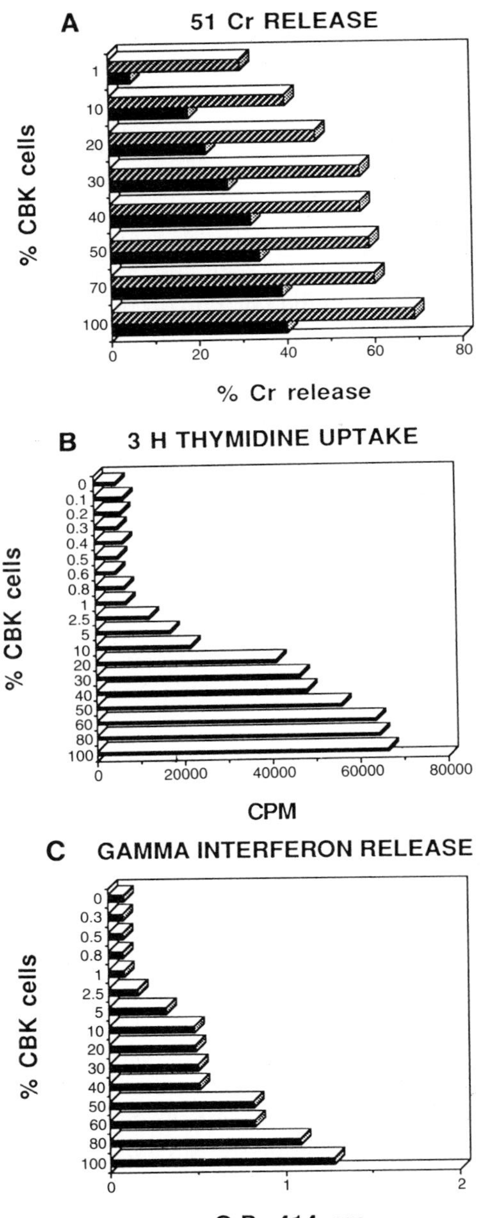

Figure 6. Response of T cells from BM3 TC3–Tg mice to H2-K^b antigen. Spleen cells from a ten-week-old BM3 responder mouse (black bars) were stimulated with mixtures of CBA and CBK (H2-K^b) cells and assayed for (A) ^{51}Cr release, (B) [^{3}H]-thymidine incorporation, or (C) γ-IFN release. Results when a younger (four to five weeks-old) BM3 mouse was used in the ^{51}Cr release assay are also presented (A, striped bars).

Protocol 9. Skin grafting

Reagents

- Anaesthetic (e.g. Hypnorm + Hypnoval)
- Sterile gauze rubbed in paraffin oil (2 cm^2)
- Plaster of Paris, 2 cm × 12 cm (Gypsona, Smith and Nephew)

Method

1. Kill graft donor mice. Wash tail with 70% alcohol. Cut around base of tail with scalpel and along the tail. Pull off tail skin and transfer to PBS in Petri dish.
2. Cut tail skin into squares of ~ 1 cm^2. Up to six grafts can be made from one donor.
3. Anaesthetize recipient mice. Shave flanks using electric hair cutter. Immobilize mice by taping fore and hind legs to a plastic tray. Wipe shaved area with 70% alcohol.
4. Cut a graft bed to receive donor skin graft. Place skin graft in bed and cover with sterile gauze.
5. Roll plaster around the belly of the mouse. Leave mice to recover (keep warm).
6. Remove casts after ten days. Check graft appearance daily. Rapid rejection manifests as scabs. Slow rejection manifests as gradual shrinking of the graft.
7. Use at least six recipients per experimental group.

In *Table 5* we present data obtained by grafting tail skin from four types of Tg mice on to CBA/Ca recipients. Grafts from CBK and KAL mice are rejected rapidly, on account of the expression of H2-K^b on keratinocytes resident in the grafts. However, grafts from Kβ and CD2K^b-3 mice were not rejected, probably because cells resident in grafts (e.g. keratinocytes) do not express H2-K^b in these Tg mice.

3.2.4 Detection of MHC transgene expression on various cells and tissues

The pattern of expression of an MHC transgene can be assessed using the techniques described above. In this section we will illustrate the versatility of these techniques by presenting examples from our own research with H2-K^b Tg mice.

i. H2-K^b expression on erythrocytes (RBCs)

By setting appropriate 'gates' using the criteria of forward (FSC) and side (SSC) light scattering characteristics during flow cytometry, RBCs and PBLs from blood samples can be readily distinguished (*Figure 7*). In this example, Kβ Tg mice, in which a β-globin promoter is linked to a H2-K^b structural gene

Table 5. Assaying transgene expression by skin grafting

Donor	Recipient	Grafts accepted /total	Mean survival time Days (range)
CBK	CBA	0/23	18 (14–22)
KAL	CBA	0/6	13
Kβ	CBA	5/5	>150
CD2K^b-3	CBA	4/4	>150

(9), are shown to express H2-K^b on RBCs but express no detectable H2-K^b on their PBLs. See notes.

ii. H2-K^b expression in thymus

The sensitivity of functional assays using T cells from TCR–Tg mice is emphasized in *Figure 8*. In this example, homogeneous cell suspensions were prepared from whole thymus of mice (after digestion in collagenase and DNase) and used to stimulate γ-IFN release by responder T cells from BM3 TCR–Tg mice. The results show that cells expressing H2-K^b are present in thymus from KAL and Kβ Tg mice. Immunofluorescent analyses, of the type shown in *Figure 3*, failed to detect H2-K^b expression in thymus from these mice.

iii. H2-K^b expression on granulocyte/macrophage (GM) cell lineages

Cells of GM lineages include dendritic cells (DCs) and macrophages. GM lineage cells can be isolated from most tissues using rather laborious procedures which require large numbers of mice (supplies of Tg mice are often limited). Culturing bone marrow *in vitro* with GM-CSF results in the differentiation of large numbers of relatively pure DCs and macrophages (*Protocol 10* and ref. 13) which can be separated on the basis of their different capacities to adhere to surfaces and according to whether they express MHC class II molecules (DCs express MHC class II constitutively). These cells are useful for functional studies as they present antigens to T cells efficiently. However, some markers expressed by these cells are not expressed by GM lineage cells found in tissues and vice versa.

Protocol 10. Preparation of dendritic cells and macrophages from bone marrow

Reagents

- Recombinant mouse GM-CSF (DNAX)
- As for *Protocol 5*

Method

1. Dissect two femurs (keep sterile). Cut off the ends of the bone. Flush through with wash medium two or three times. Resuspend and remove clumps of cells.

Protocol 10. *Continued*

2. Wash twice in wash medium. Resuspend in culture medium.
3. Culture 3×10^6 cells/ml in culture medium containing 250 U GM-CSF/ml for two days at 37 °C.
4. Remove non-adherent cells (granulocytes). Wash adherent cells gently and add fresh medium containing recombinant GM-CSF. Culture for at least another eight days.
5. Remove non-adherent cells (DCs) by pipetting them off. Remove adherent macrophages by treating trypsin/versene.

In *Figure 9*, non-adherent DCs prepared from bone marrow of various mice were used to stimulate T cells from TCR–Tg (BM3) mice (*Figure 9A*) and were subjected to cytofluorometric analyses (*Figure 9B*) to assess H2-K^b expression. This example shows that both types of experimental Tg mice express H2-K^b on DCs since γ-IFN release was detected in both cases (Kβ and CD2K^b-3) but parallel cytofluorometric analyses revealed surface H2-K^b expression in only one case (CD2K^b-3).

4. Immunological status of MHC transgenic mice

MHC Tg mice have been used extensively to investigate numerous immunological phenomena, such as tolerance induction, some forms of autoimmunity, tissue transplantation, and antigen presentation. In this section we will review their use to study factors affecting tolerance induction and xenograft rejection.

4.1 Assessing tolerance status and the mechanism of tolerance induction

T cell tolerance to self-MHC molecules is normally acquired during thymocyte development and leads to non-responsiveness both *in vitro* and *in vivo*. There are many ways of determining tolerance status experimentally. Ability to reject skin grafts expressing self-MHC molecules is one way of determining tolerance status *in vivo*. The strategy we have employed is indicated in *Figure 10*. Failure of H2-K^b Tg mice to reject skin grafts from CBK mice reveals that they are tolerant; rejection indicates responsiveness. However, this does not reveal information on the status of the T cell repertoire since a tolerant state can be induced and maintained even though potentially self-MHC reactive T cells are present. One way of demonstrating this is to backcross TCR–Tg mice

Figure 7. Cytoflourometric analyses of H2-K^b expression on RBCs (B) and PBLs (C) from Kβ Tg mice. Light scatter profiles of forward (FSC) versus side (SSC) scatter used to select RBCs (A1) and PBLs (A2) are shown. Histogrames show staining profiles of cells from C57BL stained with FITC-conjugated goat anti-mouse IgG H chain antibody alone (top left) or including anti-H2-K^b antibody (B28.24.3, top right); non-Tg littermates (CBA mice, bottom left) and Kβ Tg (bottom right).

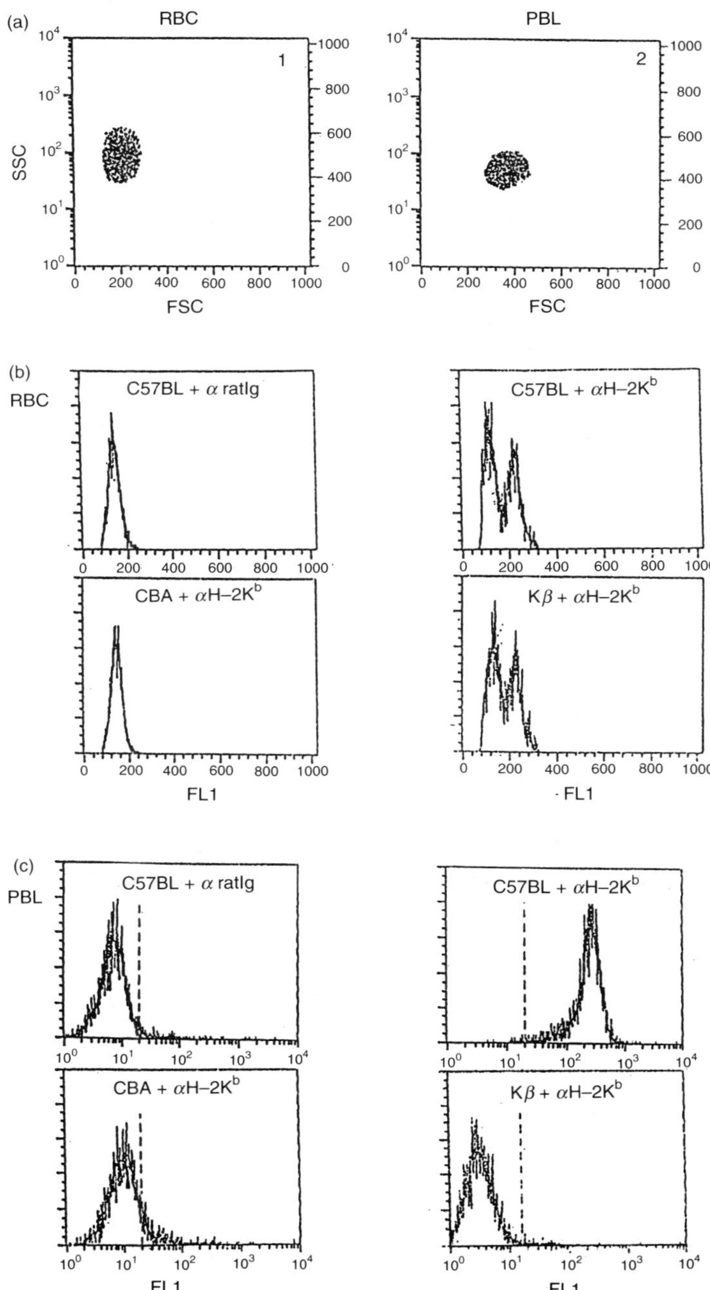
(a)
RBC
PBL
SSC
FSC
1
2
(b)
RBC
C57BL + α ratIg
C57BL + αH–2K^b
CBA + αH–2K^b
Kβ + αH–2K^b
FL1
(c)
PBL
C57BL + α ratIg
C57BL + αH–2K^b
CBA + αH–2K^b
Kβ + αH–2K^b
FL1

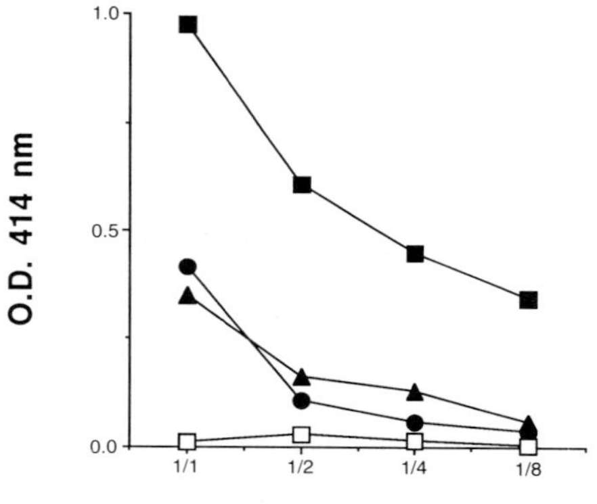

Figure 8. Expression of H2-K^b in the thymus of transgenic KAL and Kb mice. Cells from collagenase treated thymus preparations of CBK (■), KAL (●), Kβ (▲), and CBA (□) mice were used to stimulate T cells from BM3 mice. γ-IFN released into the supernatant was measured by ELISA.

with H2-K^b Tg mice and determine the fate and functional status of thymocytes and T cells in double Tg mice (*Figure 11*).

4.2 MHC Tg mice and xenotransplantation

Human organs for transplantation are in short supply and this problem is worsening as demand increases and the supply of donor organs is reduced (due in part to reductions in the number of road accidents). Therefore, there is a need to find alternative sources or organs for transplantation and recently, the possibility of using organs from other species, particularly pigs, has been discussed. The greatest barrier preventing the use of xenogeneic organ grafts is the phenomenon of hyper-acute rejection of xenotransplants mediated by the host's natural antibodies and complement. Tg pigs expressing potent inhibitors of activation of human complement (e.g. decay accelerating factor, DAF, and membrane cofactor protein, MCP or CD46) are being generated in attempts to overcome hyper-acute rejection. Problems due to T cell responses to xenoantigens expressed by organ grafts may also arise. Mouse T cells respond poorly to HLA class I antigens on transfected mouse cell lines (14). This could be explained if mouse T cells are not able to recognize human MHC class I antigens because of structural incompatibility of TCR–MHC interactions or if co-receptor (CD4 and CD8)/MHC interactions are suboptimal.

Figure 9. H2-K^b expression on GM-CSF induced non-adherent cells (dendritic cells) from H2-K^b Tg mice. Non-adherent cells from B10 (■), KAL (●), Kβ (▲), CD2K^b-3 (+), and CBA (□) mice were used to stimulate speel cells from BM3 TCR–Tg mice. γ-IFN released into supernatants was measured. Histograms show the results of cytofluorometric analyses of the same cells stained with biotinylated anti-H2-K^b antibody (MV3) and developed with streptavidin–PE.

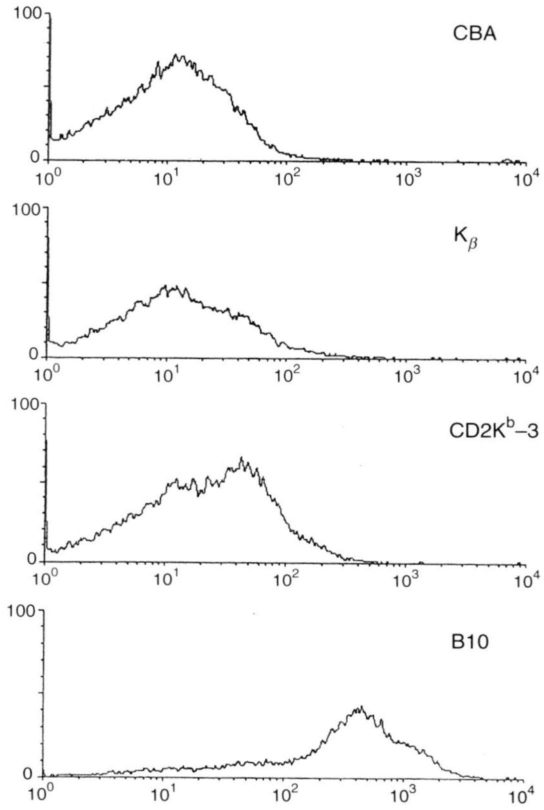
100
0
CBA
10^0
10^1
10^2
10^3
10^4
K_β
$CD2K^b$–3
B10

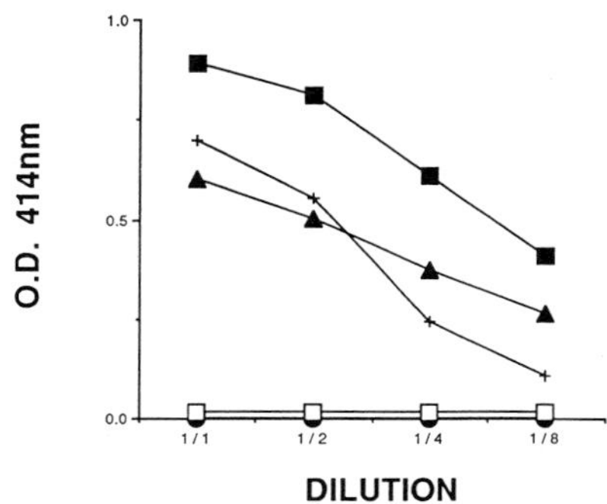
O.D. 414nm
1.0
0.5
0.0
1/1
1/2
1/4
1/8
DILUTION

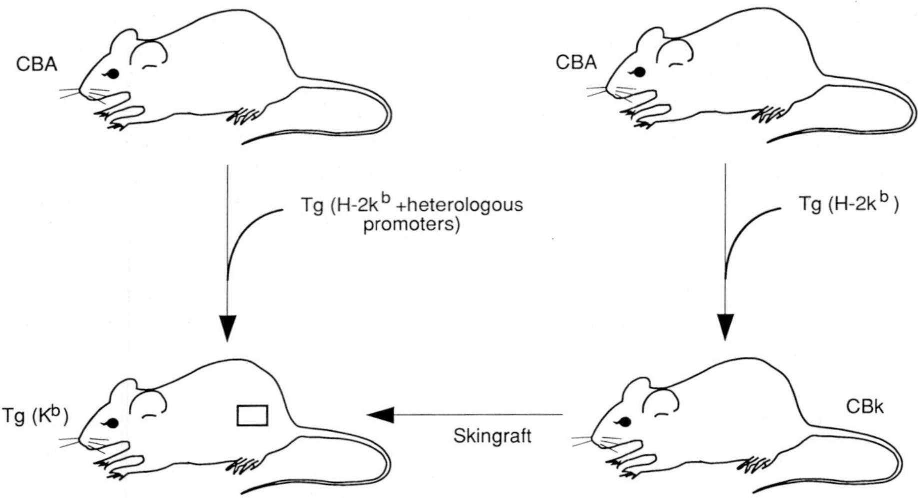

Figure 10. Testing for tolerance by skin grafting.

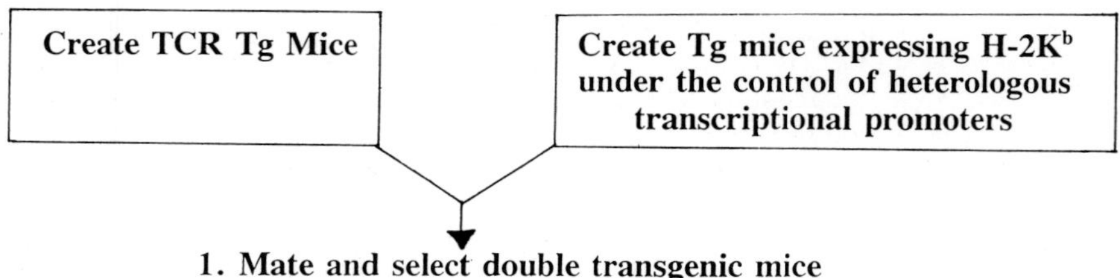

1. Mate and select double transgenic mice
2. Determine the fate of TCR+ cells by cytofluorimetry.
3. Test peripheral T cells for function

Figure 11. Strategy for determining the mechanism of tolerance induciton using TCR–Tg mice.

Human MHC class I (HLA) Tg mice have been generated to assess T cell responses across xenogeneic barriers. HLA-A, -B, and -C class I Tg mice have been studied, but there is conflicting evidence on the ability of naïve mice to generate cytotoxicity directed against HLA class I antigens. In one report, using both HLA-B27 and HLA-Cw3 transgenic mice, anti-HLA (xeno) reactivity was about 200 fold weaker than anti-H2 (allo) responses. This was ascribed to the failure of mouse CD8 to recognize the α3 domain of the HLA molecule and was not due to a failure to generate an anti-HLA TCR repertoire (15). Conflicting results were reported for the same (16) and for other HLA Tg mice (17). In these studies, anti-xeno and anti-allo T cell precursor frequencies were similar. Despite their failure to provoke primary T cell responses *in vitro*, transgenic HLA class I antigens did provoke rapid skin graft rejection (16). Secondary T cell responses to transgenic HLA antigens were also detected and, in some cases, HLA antigens were recognized in association with H2 antigens (18). These studies demonstrate that there is cross-species recognition of MHC antigens by T cells. This must be taken into consideration when considering xenotransplantation.

References

1 Hogan, B., Constantini, F., and Lacy, E. (1986). *Manipulating the mouse embryo.* Cold Spring Harbor Press, Cold Spring Harbor, New York.
2. Murphy, D. and Carter, D. A. (1993). *Transgenesis techniques: principles and protocols.* Humana Press, Totowa, New Jersey.
3. Yeoman, H. and Mellor, A. L. (1992). *Int. Immunol.*, **4**, 59.
4. Mamalaki, C., Tanaka, Y., Corbella, P., Chandler, P., Simpson, E., and Kioussis, D. (1993). *Int. Immunol.*, **5**, 1285.
5. Bander, S. A. A., Watson, S. C., and Shire, J. G. M. (1988). *J. Reprod. Fertil.*, **84**, 133.
6. Maniatis, T., Fritsch, E. F., and Sambrook, J. (ed.) (1982). *Molecular cloning, a laboratory manual.* Cold Spring Harbor Press, Cold Spring Harbor, New York.
7. Husbands, S. D., Schönrich, G., Arnold, B., Chandler, P. R., Simpson, E., Philpott, K. L., *et al.* (1992). *Eur. J. Immunol.*, **22**, 2655.
8. Simpson, S. J., Tomlinson, P. D., and Mellor, A. L. (1993). *Int. Immunol.*, **5**, 189.
9. Sponaas, A.-M., Tomlinson, P. D., Antoniou, J., Auphan, N., Langlet, C., Malissen, B., *et al.* (1994). *Int. Immunol.*, **6**, 277.
10. Prat, M., Gribaudo, G., Comoglio, M., Cavalio, G., and Landolfo, S. (1984). *Proc. Natl. Acad. Sci. USA*, **81**, 4515.
11. Simpson, E. and Chandler, P. (1986). In *Handbook of experimental immunology* (ed. D. M. Weir, L. A. Herzenberg, and L. A. Herzenberg), Vol. 2, Ch. 68. Blackwell Scientific, Oxford.
12. Billingham, R. E. and Medawar, P. B. (1951). *J. Exp. Biol.*, **28**, 385.
13. Inaba, K., Inaba, M., Romani, N., Aya, H., Deguchi, M., Ikehara, S., *et al.* (1992). *J. Exp. Med.*, **176**, 1693.
14. Moots, R. J., Samberg, N. L., Pazmany, L., Frelinger, J. A., McMichael, A. J., and Stauss, H. J. (1992). *Eur. J. Immunol.*, **22**, 1643.
15. Kalinke, U., Arnold, B., and Hämmerling, G. (1990). *Nature*, **348**, 642.
16. Kievits, F., Boerenkamp, W., Lokhorst, W., and Ivanyi, P. (1990). *J. Immunol.*, **144**, 4513.
17. Epstein, H., Hardy, R., May, J. S., Johnson, M. H., and Holmes, N. (1989). *Eur. J. Immunol.*, **19**, 1575.
18. Dill, O., Kievitz, F., Koch, P., Ivanyi, P., and Hämmerling, G. (1988). *Proc. Natl. Acad. Sci. USA*, **85**, 5665.

7

Imaging of individual cell surface MHC antigens using fluorescent particles

PATRICIA R. SMITH, KEITH M. WILSON, IAN E. G. MORRISON, RICHARD J. CHERRY, and NELSON FERNANDEZ

1. Introduction

Class I and class II major histocompatibility complex (MHC) molecules are cell surface, transmembrane glycoproteins that have similar overall structure but have readily distinguishable subunit structures (1). The two classes of MHC molecules present peptides, produced from different intracellular compartments, at the cell surface where they are recognized by two major subsets of T cells, CD4 and CD8, respectively.

Integral membrane glycoproteins are able to diffuse to varying degrees within the plane of the membrane and so can potentially form protein clusters (2). This clustering of cell surface receptors in response to a particular signal or antibody is important in transmembrane signal transduction (3, 4). The density of class I and class II subsets and their relationship with other cell surface molecules in the correct orientation is thought to be vital for effective T cell recognition. For example, it has been shown that human class I molecules cluster in liposomes and on the surface of JY lymphoblasts (5), normal B and T cells, cells of B and T lymphoblast lines, and transformed fibroblasts (6). In addition, dimers and trimers of mouse class I molecules have also been detected in transformed cells (7). The relationship of antigen density and function has been shown in the case of allogenic T cell clones whose suppressor properties correlate with a high epitope density of class II molecules (8). Additionally, heterotypic association was shown in the case of human class II products which form oligomeric compounds with a number of co-precipitating proteins (9, 10). A high-resolution crystal structure of the human HLA class II DR1 molecule suggests that the protein crystallizes as a dimer of dimers (11); highlighting that a similar functional unit in the membrane could act by cross-linking T cell receptors. Although there is no direct evidence for the existence of such structures in the native cell membrane, these studies suggest the importance of association of class II molecules in cell signalling (12, 13).

A major limitation in the analysis of MHC lateral mobility and clustering has been the availability of suitable techniques. The techniques most frequently employed rely on fluorescence microscopy, in particular the method of photobleaching of fluorescent probes known as FRAP (fluorescence recovery after photobleaching) (14, 15). Here, fluorescent molecules within an area of the cell surface are photobleached by a brief laser pulse. Measurement of recovery of fluorescence in that area indicates the rate of diffusion of unbleached fluorophores into the illuminated area. Using this technique it has been shown that the lateral mobility of most membrane proteins is highly restricted. The main disadvantage of FRAP is that it measures the average properties of a large number of molecules and only distinguishes between mobile and immobile particles. This technique is also poor at discriminating between different types of motion. More importantly, however, it can only measure diffusion over distances of the order of 1 μm, so failing to detect the mobility of molecules which are constrained to move over submicrometre distances.

An alternative technique for studying diffusion involves the use of a sensitive imaging system to detect low intensity fluorescent emission. In this way the paths of small fluorescent particles can be tracked over a period of time. This largely overcomes the problems associated with FRAP and allows a more detailed analysis of membrane protein mobility. The feasibility of tracking individual receptors was first demonstrated using low density lipoproteins (LDL) labelled with about 40 fluorescent probes and visualized with an image-intensified video camera (16, 17). A current system based on a cooled, slow-scan, charge-coupled device (CCD) camera for measuring motions of individual particles by fluorescence digital imaging, can locate particles to within 25 nm. This type of imaging system has since been semi-automated so that particles can be traced from frame to frame among even quite densely packed surface areas, and has been used to demonstrate the very low but measurable mobility of LDL receptors at 4 °C (18).

In order to apply this approach to immunological receptors such as cell surface MHC molecules, fluorescent particles need to be attached to a suitable biomolecule which will recognize the MHC antigen of interest. In our laboratory, we have recently shown that phycobiliproteins coupled to monoclonal antibodies provide a sufficiently bright signal for an individual fluorescent particle to be detected (see Section 3). Together with the improved imaging systems the newly available fluorophores and monoclonal antibodies combine to produce an extremely powerful tool for investigating the lateral mobility of individual MHC class I and class II molecules on the cell surface.

2. Preparation of Fab fragments from ascites fluid and hybridoma supernatant

In order to image individual MHC antigen subsets at the cell surface it is necessary to select a suitable monoclonal antibody with high affinity for a given epitope. The antibody is then conjugated to a fluorescent probe, e.g. phycobiliproteins. The monoclonal antibody will either be in the form of ascites fluid (1–10 mg/ml) or hybridoma supernatant (5–40 μg/ml). The IgG has to be purified from these fluids. Ascites fluid often contains precipitates and lipids which need to be removed by high speed centrifugation. However, IgG often associates with the lipid layer; washing of this lipid layer will often enhance the recovery of IgG. After centrifugation the IgG can be isolated from the ascites fluid using an affinity column made of protein A or protein G depending on the IgG isotype of the monoclonal antibody. Affinity matrices of this type are available in loose form or pre-packed in columns. Although more expensive, the pre-packed columns are much more convenient and labour-saving, and are usually more efficient. If the hybridomas are cultured in the presence of serum, there will be bovine IgG present in the supernatant which will also bind to the protein A or G. This IgG must be considered when calculating the volume of supernatant to be loaded onto the affinity column. The IgG will elute off the affinity column in a relatively small volume (1–2 ml) but this may still need concentrating further. This is achieved using centrifugeable ultrafiltration devices with a molecular weight cut-off of 30 kDa (Amicon). The IgG can be concentrated in these devices until the required concentration is achieved.

Protocol 1. Purification of IgG from ascites fluid or hybridoma supernatant

Equipment

- High speed refrigerated centrifuge
- Bench-top refrigerated centrifuge
- Fraction collector, peristaltic pump, UV monitor, plotter (e.g. Pharmacia)
- UV/vis spectrophotometer

A. *Ascites fluid*

1. Centrifuge the ascites fluid (200 *g*, 20 min (4°C). Aspirate off the top lipid layer and retain. Also aspirate off the supernatant below this containing the clean fluid. Discard the pellet.
2. Wash the fatty layer twice with phosphate buffer (20 mM, pH 7.0, 1 ml) by centrifugation as above.
3. Combine the supernatant and the lipid layer washes and load on to a 1 ml protein A or protein G HiTrap column[a] (Pharmacia, 20 mg

Protocol 1. *Continued*

capacity) at 0.5 ml/min, washing through with phosphate buffer (20 mM, pH 7.0, 10 column volumes). Discard the unbound washings.

4. Elute the bound IgG with citric acid (0.1 M, pH 3.0), into microtitre tubes containing Tris–HCl (1 M, pH 9.0, approx. 50 μl/ml).
5. Combine the protein-containing fractions, check the pH, and if necessary adjust using Tris–HCl 1 M, pH 9.0).
6. Calculate the IgG concentration as follows:

$$\varepsilon_{(mM)} = 1.34 \text{ at } 280 \text{ nm}$$

$$\text{Abs} = \varepsilon_{(mM)} \times c \times l$$

where ε = extinction coefficient of mouse IgG at 280 nm, c = concentration, and l = path length (usually 1).

B. *Hybridoma supernatant*

1. Load on to a 1 ml protein A or protein G HiTrap column[a] (Pharmacia) at 0.5 ml/min, washing through with phosphate buffer (20 mM, pH 7.0, 10 column volumes). Discard the unbound washings.
2. Elute the bound IgG from the affinity column with citric acid (0.1 M, pH 3.0), into microtitre tubes containing Tris–HCl (1 M, pH 9.0, approx. 50 μl/ml).
3. Combine the protein-containing fractions, check the pH, and if necessary adjust using Tris–HCl (1 M, pH 9.0).
4. Calculate the IgG concentration as shown in part A.

[a] Use protein A or protein G depending on the IgG isotype of the monoclonal antibody.

An IgG molecule can in principle bind two molecules of antigen, and this divalency may affect the analysis and interpretation of images obtained. To overcome this problem, the use of monovalent Fab fragments generated from the IgG molecules is advisable. These fragments retain antigen specificity whilst exhibiting low non-specific binding. The use of Fab fragments to bind cell surface antigens also greatly reduces the chances of spontaneous capping or aggregation of antigen following binding with antibody (19).

Papain is the enzyme commonly used to digest IgG into Fab and Fc fragments. Papain contains an active cysteine, the sulfydryl group of which must be in the reduced form for activity. In the presence of cysteine, papain splits the IgG molecule at one or more disulfide bonds in the hinge region, producing two uncoupled Fab fragments and one Fc fragment. The papain concentration and the digestion time required varies among the different IgG subclasses. For example, IgG1 is most sensitive to proteolysis followed by IgG2a > IgG3 > IgG2b. The sensitivity to proteolysis also varies from one monoclonal antibody to another. It is, therefore, necessary, to carry out a range of small scale digests in order to determine the optimum conditions. In most cases, varying the digestion time whilst keeping constant papain and cysteine concentrations and constant pH will suffice. With some monoclonal antibodies, however, it may be necessary to alter the papain, cysteine, and pH in addition to the time. After digestion, it is necessary to inactivate the cysteine, as well as the papain, by the addition of iodoacetamide. Without this step, the cysteine will reduce internal Fab disulfide bonds eventually leading to the formation of Fab aggregates. After digestion the Fc and any undigested IgG can be removed using a protein A or protein G column. The purity of the Fab preparation can be checked by size exclusion HPLC (*Figure 1*) or if this is not available by SDS–PAGE. Fab is not a particular stable protein and has a tendency to form high molecular weight aggregates. The addition of salt (0.1 M NaCl) to the digest helps to prevent aggregate formation. After column chromatography it is usually necessary to concentrate the Fab and remove the iodoacetamide and cysteine–HCl by dialysis. Fab does not behave as a globular protein with a molecular weight of 50 kDa and a significant amount of it will pass through a dialysis membrane with a molecular weight cut-off of 30 kDa. A

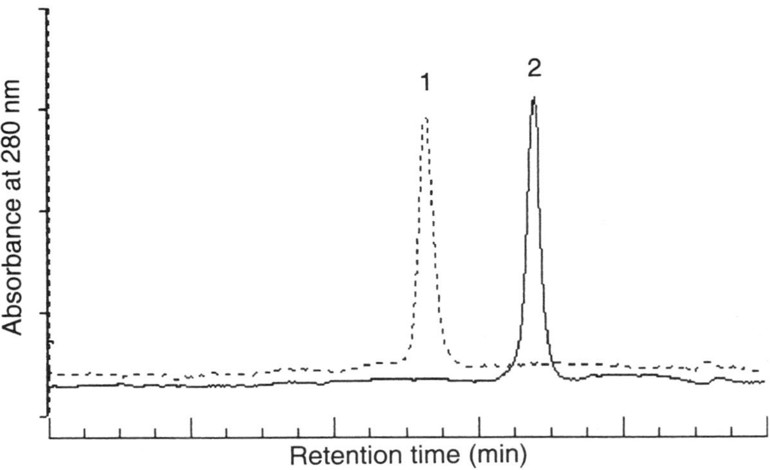

Figure 1. Analytical HPLC size exclusion chromatography of affinity purified IgG and Fab fragments. Aliquots of affinity purified IgG (dotted line) and Fab fragments (solid line) were loaded onto a Bio-Select SEC 250-5 column (Bio-Rad) and eluted with 50 mM sodium phosphate, 150 mM sodium chloride pH 6.8 at 1.0 ml/min. IgG elutes at a position corresponding to a M_r of 153 kDa (1) and Fab elutes at a position corresponding to a M_r of 30 kDa (2).

significant amount of Fab will also precipitate on to certain filtration materials. Use of a cellulose acetate ultrafiltration membrane with a 5 kDa molecular weight exclusion limit (Sartorius) ensures that the Fab does not precipitate on to the membrane or pass through it. Size exclusion HPLC can be used to monitor aggregate formation during concentration and dialysis, and also to remove aggregates prior to conjugation to a fluorophore.

Protocol 2. Preparation of Fab fragments from IgG

Equipment

- 37 °C shaking incubator
- Bench-top microcentrifuge
- Refrigerated bench-top centrifuge
- UV/vis spectrophotometer
- Size exclusion HLPC or SDS—PAGE

Method

1. Concentrate the IgG solution to approx. 20 mg/ml using centricon 30 or centriprep 30 concentrators (Amicon). Use the concentrators to dialyse the IgG into phosphate buffer (20 mM, pH 7.0) containing EDTA (10 mM).
2. On the day of use prepare the digestion buffer by dissolving cysteine–HCl (42 mg) in phosphate buffer (12 ml, 20 mM, pH 10). The final pH should be 7.0 ± 0.2
3. Microcentrifuge immobilized papain slurry (0.5 ml) (Pierce) in a 2 ml microtitre tube. Discard the supernatant and wash the papain pellet with 3 × 2 ml aliquots of digestion buffer. Resuspend the pellet in digestion buffer (0.5 ml).
4. Add digestion buffer (0.5 ml) to the IgG solution (0.5 ml, 20 mg/ml) and combine with the papain solution. Incubate in a 37 °C shaking incubator for 5–24 h. (It must be shaking vigorously to prevent settling of the papain.)
5. Add iodoacetamide (2.8 mg/ml) to stop the digestion. Put on a whirly wheel at room temperature for 15 min.
6. Microcentrifuge the digestion solution and collect the supernatant. Wash the papain pellet with 2 × 0.5 ml aliquots of phosphate buffer (20 mM, pH 7.0) containing NaCl (0.1 M). Combine the washings and the supernatant.
7. Load the IgG digest on to a protein A or protein G 1 ml HiTrap affinity column[a] at 0.5 ml/min, washing through with phosphate buffer (20 mM, pH 7.0) containing NaCl (0.1 M). Collect the unbound Fab in 0.5 ml fractions.

8. Combine the Fab fractions and remove the iodoacetamide by centrifugal ultrafiltration using a centristart 1 device (5 kDa M_r exclusion limit) (Sartorius). This also serves to concentrate the Fab.
9. Check the purity of the Fab by either SDS–PAGE or size exclusion HPLC, using for example, a 250 Bioselect column (Bio-Rad).
10. If Fab aggregates are apparent, these can be removed by applying the Fab to the size exclusion HPLC column and collecting the correct molecular weight fractions. Maintain the salt content (0.1 M).
11. Calculate the Fab concentration as shown in *Protocol 1A*.
12. The bound IgG and Fc can be eluted with citric acid (0.1 M, pH 3.0) and any undigested IgG can be detected by size exclusion HPLC and recovered for future use after dialysis back to pH 7.0.

[a] Use protein A or G depending on the IgG isotype of the monoclonal antibody.

After digestion and prior to conjugation to a fluorophore it is necessary to check that the Fab has retained its specificity. This is done by indirect immunofluorescence using the appropriate non-adherent cell line; e.g. for human MHC reactivity use a lymphoblastoid B cell line (see Section 4 for an alternative method). This is an analytical technique where the reaction of the Fab can be compared to that of the original ascites fluid or cell culture supernatant, whole IgG, and a negative control.

Protocol 3. Indirect immunofluorescence of Fab fragments

Equipment

- Fluorescence microscope
- Bench-top centrifuge

Method

1. Isolate 200 000–500 000 cells/test sample from an appropriate cell line.
2. Centrifuge the cells (3000 r.p.m., 15 min) and wash twice with indirect immunofluorescence buffer (IF) (PBS, 1% BSA, 0.02% NaN_3).
3. Resuspend the cell pellet in IF buffer (50 μl/test sample) and aliquot into the required number of microtitre tubes.
4. Microcentrifuge each tube and resuspend the pellets in the required antibody (50 μl) diluted in IF buffer (1/100 for IgG). Put the tubes on a whirly wheel for 30 min at room temperature.
5. Microcentrifuge the tubes and wash three times with IF buffer (1 ml).

Protocol 3. *Continued*

6. Add fluorescein-conjugated rabbit anti-mouse (1/50 dilution, 50 μl) to each pellet and incubate on a whirly wheel at room temperature for 30 min.
7. Microcentrifuge the tubes and wash three times with IF buffer (1 ml).
8. Add 10 μl IF buffer to each pellet and vortex before fixing with paraformaldehyde (4%, 400 μl). This stops the cells clumping.
9. If the cells are to be stored overnight add PBS/DABCO (diazabicyclo (2.2.2)octane) (10%, 10 μl). This helps to prevent photobleaching.
10. Mount the cells on microscope slides and seal the coverslips with silicone grease to prevent drying out.
11. Examine the slides using a fluorescence microscope containing the appropriate filters. Cells containing bound fluorescein-conjugated antibody will fluoresce green.

3. Preparation of phycobiliprotein–Fab conjugates

Phycobiliproteins are stable, highly soluble proteins derived from cyanobacteria and eukaryotic algae which contain a monodisperse population of prosthetic fluorophores (20). Because of their biological role as light harvesters requiring maximal absorbance and fluorescence, they have extremely good absorbance and fluorescence characteristics. Five- to tenfold increases in sensitivity have been reported with the B–phycoerythrin antibody conjugate compared to the corresponding fluorescein conjugate. Several different phycobiliproteins are available commercially with different excitation and emission wavelengths. The type of excitation laser available determines the particular phycobiliprotein required (*Table 1*, adapted from the Molecular Probes Handbook).

For immunological studies the phycobiliproteins are conjugated to other proteins, in this case Fab fragments generated from an MHC-specific monoclonal antibody. This can be achieved through the use of non-covalent reactions, such as that of biotin and streptavidin. It is always better, however, to produce a stable covalently bound conjugate. There are several established methods for conjugating one protein to another, e.g. *N,N′-o*-phenylenedimaleimide, glutaraldehyde, and *m*-maleimideobenzoyl-*N*-hydroxysuccinimide ester (21). The major manufacturer of phycobiliproteins (Molecular Probes) recommend producing a thioether protein–protein linkage where the phycobiliprotein is converted to a pyridydisulfide derivative, which can then be reduced to the active thiol as required. We recommend that phycobiliproteins are purchased as the ready made pyridyldisulfide derivatives. This eliminates one step in an already lengthy process, the derivative comes under quality control checks, and it prevents light exposure and dilution of the

Table 1. Excitation and emission wavelengths of phycobiliproteins[a]

Source	Phycobiliprotein	λ_{exc} (nm)	ε_{max} (cm^{-1} M^{-1}) (at λ_{max} (nm)	λ_{em} (nm)
Argon ion laser	R–phycoerythrin (R–PE)	515	1 960 000 (565)	578
Argon ion laser	B–phycoerythrin (B–PE)	568	2 410 000 (565)	575
Helium–neon laser	Allophycocyanin	632	700 000 (650)	660

[a] λ_{exc} = maximum excitation wavelength, ε_{max} = maximum molar extinction coefficient, λ_{em} = maximum fluorescence wavelength.

phycobiliproteins. We have found that the phycobiliproteins tend to deteriorate with time (after approximately six months), therefore, any steps which reduce the risk of waste or photobleaching need to be utilized. The lysines of the antibody to be conjugated are easily converted to thiol reactive maleimides with a heterobifunctional cross-linking reagent, such as SMCC (4-(maleimidomethyl)cyclohexanecarboxylic acid *N*-hydroxysuccinimide ester).

The relative amounts of Fab and phycobiliprotein used depends on the imaging experiment envisaged. If individual phycobiliprotein antigens are to be imaged then a 1:1, Fab:phycobiliprotein ratio is required. The number of pyridyldisulfides per phycobiliprotein varies from batch to batch, usually 1.9–2.4 residues per phycobiliprotein. This is only an average figure, and so some phycobiliproteins may have ten derivatives and some may have none at all. In our experience it is better to have a lower rather than higher average, and if possible to request a sample from a lower average batch. The amount of SMCC used to derivatize the Fab determines the degree of derivatization of the Fab. The more lysines that are derivatized on an individual Fab molecule, the more phycobiliprotein molecules it is likely to react with. This will serve to increase the Fab:phycobiliprotein ratio. A five to ten molar excess of SMCC should ensure that a sufficient but not excessive number of lysines are derivatized per Fab molecule. In addition to this a 1:1 molar ratio of Fab:phycobiliprotein needs to be used.

Protocol 4 Preparation of a phycobiliprotein–Fab conjugate

Equipment

- Cold room

Method

1. Add solid DTT (4 mg, 50 mM) to phycobiliprotein (1 mg) in a microcentrifuge tube and incubate on a whirly wheel for 15 min at room temperature.

Protocol 4. *Continued*

2. Prepare a stock solution of SMCC (5 mM) in DMSO and add five to ten molar equivalents to a solution of Fab (5 mg/ml, 208 μg). Incubate on a whirly wheel for 30 min at room temperature.
3. Dialyse both solutions separately against phosphate buffer (20 mM, pH 7.0) containing NaCl (0.1 M), at 4°C for 24 h with at least five changes of buffer.
4. Mix the two solutions and incubate on a whirly wheel for 16–20 h at 4°C.
5. Add a 20-fold excess of *N*-ethylmaleimide to the Fab–phycobiliprotein conjugate and store at 4°C. Do not freeze this reagent.

The protocol can be scaled up or down depending on the amount of Fab available and financial limitations on the use of phycobiliproteins. We have, found, however, that using less than 1 mg of phycobiliprotein creates handling and dilution problems. Also, after the conjugate has been purified and tested for reactivity the amount left for actual imaging is reduced greatly. Once the crude phycobiliprotein–Fab conjugate has been prepared it is desirable to separate out the different species. The main problem in doing this is that one Fab attached to one phycobiliprotein (240 kDa molecular weight) does not significantly alter the physical and chemical characteristics of the phycobiliprotein.

We have found that size exclusion chromatography provides a reasonable separation of species. With normal pressure size exclusion chromatography the different species are quite well separated but the fractions are so dilute as to be virtually unusable. This problem can be overcome by using HPLC. Even here, however, the Fab presents a problem. During size exclusion chromatography it does not behave as a globular protein with a molecular weight of 50 kDa, but elutes off after ovalbumin (43 kDa) behaving as if it were only 30 kDa in molecular weight. Consequently, it is very difficult to separate free phycobiliprotein from the 1:1, Fab:phycobiliprotein conjugate. The HPLC profile of the crude conjugate shows large amounts of unconjugated Fab and phycobiliprotein (*Figure 2*). The 1:1 Fab:phycobiliprotein fraction is clear as a shoulder on the free phycobiliprotein peak. The later peaks contain 2:1, 3:1, etc. ratios of the conjugate. Leaving the reaction to carry on further than 20 h does not result in any further conjugation. The unreacted phycobiliprotein probably represents that which has not been derivatized with the pyridyldisulfide. The free Fab is very well separated from the phycobiliprotein and the conjugates and so can be rescued for further use. The Bio-Rad 250 Bioselect column is suitable for use as a semi-preparative column (1–5 mg) as well as an analytical column and of the columns we have tested, this one provides the best separation. It is virtually impossible to remove all the free phycobiliprotein from the probe, but the free phycobiliprotein does not

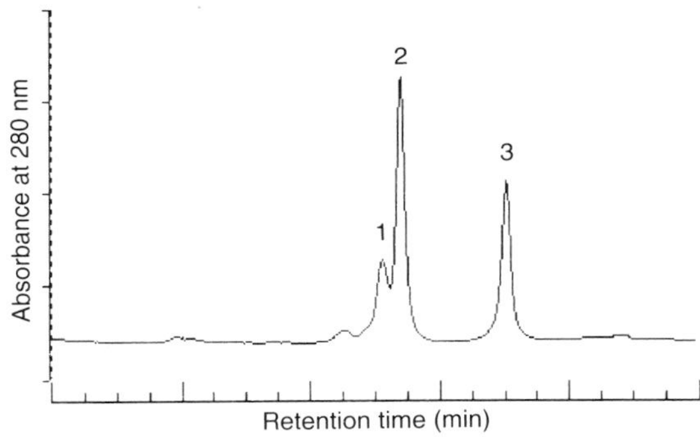

Figure 2. Purification of HLA-DR specific R–phycoerythrin labelled Fab fragments by HPLC size exclusion chromatography. Labelled Fab fragments were loaded onto a Bio-Select SEC 250-5 column (Bio-Rad) and eluted with 50 mM sodium phosphate, 150 mM sodium chloride pH 6.8 at 0.1 ml/min. The 1:1 Fab–phycoerythrin conjuage (1) is eluted close to free R–phycoerythrin (2), followed by well-separted unlabelled Fab (3).

appear to bind at all to cells. However, this needs to be verified for each individual system. The probe is susceptible to bacterial contamination and it is advisable to sterile filter it after HPLC and use under aseptic conditions thereafter. The addition of the probe to cells prohibits the use of anti-bacterial agents such as sodium azide.

Protocol 5 Purification of Fab–phycobiliprotein by size exclusion chromatography

Equipment

- HPLC with fraction collector and size exclusion column (Bio-Rad 250 Bioselect)
- 0.4 μm aqueous filter system

Method

1. Prepare the HPLC buffers Na_2SO_4 (50 mM), Na phosphate (10 mM, pH 6.8). Filter them through a 0.4 μm filter. The buffers must be filtered each day.
2. Filter the crude Fab–phycobiliprotein through an HPLC compatible 0.45 μm Acrodisc (4 mm) (Gelman).
3. Load 300 μl of sample on to the size exclusion column at 0.1 ml/min and collect 200 μl fractions into sterile microcentrifuge tubes.
4. Repeat step 3 until all of the crude conjugate has been purified.
5. Store the fractions at 4°C for further analysis and imaging. Always remove aliquots under aseptic conditions.

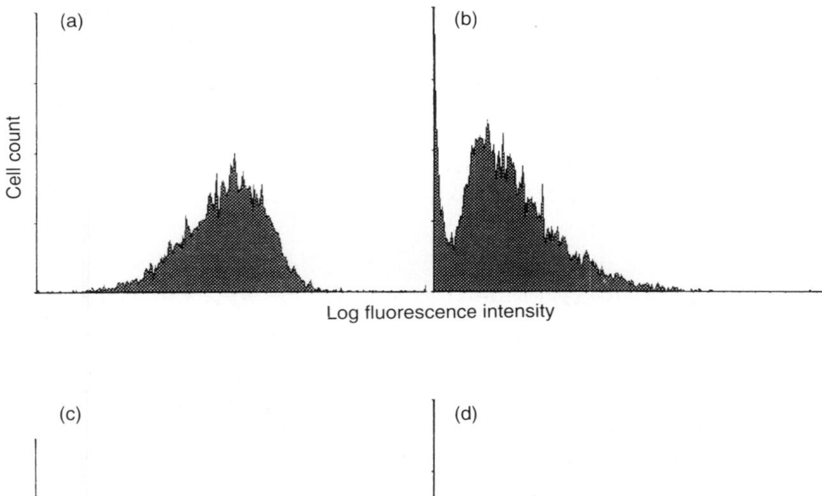

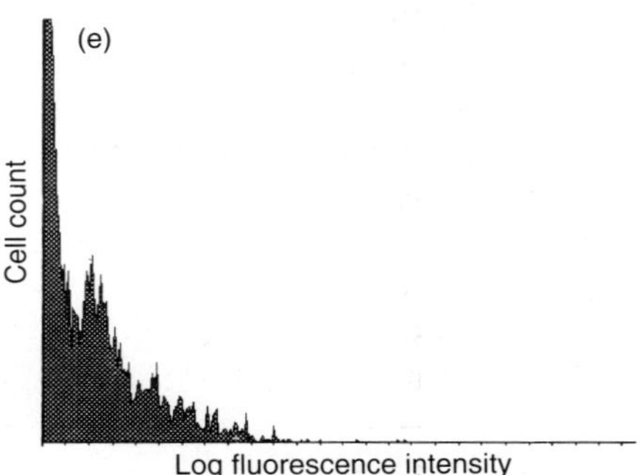

Figure 3. Flow cytometric analysis of R–phycoerythrin (PE) labelled IgG and Fab fragments binding to human class II DR receptors on a human B-LCL. Cells were incubated with 2 pmol of PE labelled IgG (A), 2 pmol of PE labelled IgG in the presence of a tenfold excess of the unlabelled IgG (B), 2 pmol of PE labelled Fab (D), and 2 pmol of PE labelled Fab in the presence of a tenfold excess of the unlabelled IgG (E). Cells were also incubated with unconjugated PE (C). 10 000 labelled cells were analysed for fluorescence on an EPICS CS flow cytometer. The histograms display relative cell numbers (*y* axis) as a function of relative fluorescence intensities (*x* axis).

If the imaging experiments envisaged do not specifically require the detection of individual cell surface molecules with individual molecules of probe, many of the above protocols are not necessary. For example, it may be possible to use whole IgG rather than Fab, or excess Fab could be conjugated with the phycobiliprotein, or the crude conjugate need not be purified by HPLC. It would still be necessary to remove any unbound Fab, but this is relatively straightforward by HPLC. Any probe produced in this way, however, would have limited uses.

4 Analysis of fluorescent probes by flow cytometry

After size exclusion chromatography of the crude Fab–phycobiliprotein conjugate the individual fractions have to be analysed for specific reactivity towards the MHC antigen of interest. A quick way to look at binding of a fluorescent probe to cells is by flow cytometry (22). Flow cytometry allows rapid measurements to be made on individual cells as they flow in a fluid stream, one by one through a sensing point. Most importantly, the results do not give an average value for the whole population but measurements are made separately on each particle within the suspension. Information yielded includes the percentage and distribution of cells with fluorescent Fab–phycobiliprotein probe bound, the different cell subpopulations present, and the effect of the probe on cell size and shape. Different cell lines can be used to determine the specificity of the probe for a particular MHC antigen, and excess Fab can be used to inhibit that specific binding (*Figure 3*). The type and number of cells used and the amount of probe required varies from probe to probe. In general, however, it is probably best to use as few cells as possible so that more probe is bound to each cell and it is therefore easier to distinguish a positive from a negative result. For analysis of human MHC antigens a lymphoblastoid cell line could be used, for example. Dead cells must be removed prior to flow cytometry as they will bind the probe non-specifically. The amount of probe required, if produced by the described protocols, will be between 10–50 μl. Using this technique the conditions required to prevent non-specific binding of probe can also be determined. For example, it may be necessary to add varying concentrations of BSA or Tween to the probe upon incubation with the cells. It is preferable to determine the specific conditions required before starting any fluorescence imaging experiments.

Protocol 6. Preparation of cells for analysis of fluorescent Fab probe by flow cytometry

Equipment

- Flow cytometer
- Bench-top centrifuge

Method

1. Count out the number of cells required (approx. 50 000 test sample) plus 50% extra to account for the removal of dead cells.

Protocol 6. *Continued*

2. Centrifuge the cells (300 *g*, 15 min, room temperature) and resuspend in PBS (2 ml).
3. Pipette Nycoprep (1.077, 2 ml) (NYCOMED) into a 10 ml test-tube and overlay the cells gently on top of it.
4. Centrifuge (1800 r.p.m., 20 min), and remove the buffy coat containing the cells by aspirating with a Pasteur pipette. Count the cells again.
5. Centrifuge the cells (3000 r.p.m., 15 min) and resuspend in immuno-fluorescence buffer (IF) (PBS, 1% BSA, 0.02% NaN_3) (50 μl, 50 000 cells per test sample).
6. Microcentrifuge the cells and resuspend in IF buffer with added probe (100 μl total volume). Incubate on a whirly wheel for 30 min at room temperature.
7. Microcentrifuge the samples and wash three times with IF buffer (1 ml).
8. Resuspend the cells in PBS (100 μl), ready for flow cytometry.

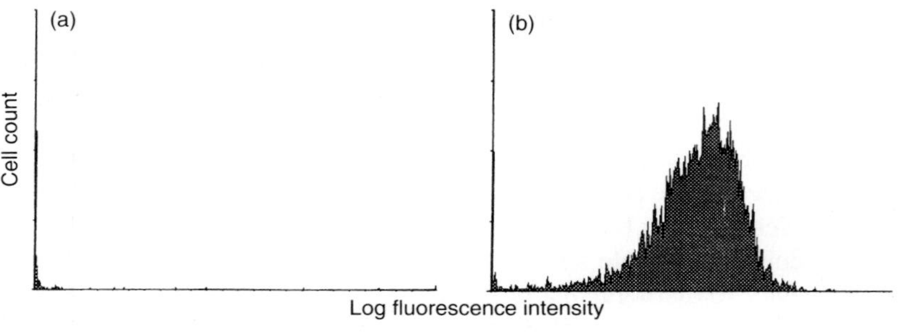

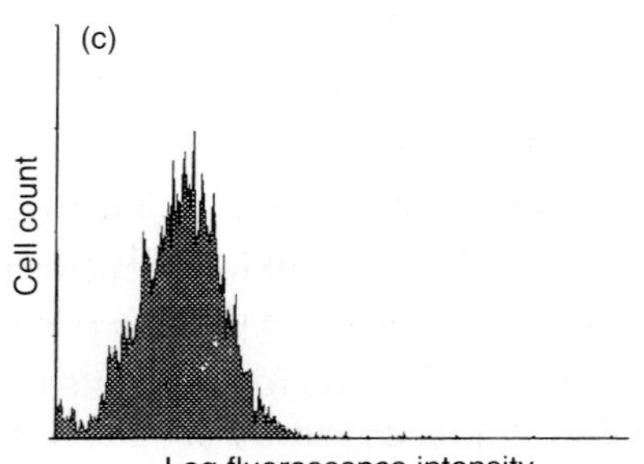

Figure 4. Flow cytometric analysis of affinity purified IgG and Fab binding to human B lymphoblastoid cells. Cells were incubagted with 15.5.5 (H2-K specific) IgG (A), HB55 (HLA-DR specific) IgG (B), and HB55 Fab (C). Cell staining was developed using an appropriate dilution of a rabbit anti-mouse FITC conjugate and 10 000 cells analysed for flouorescence on an EPICS CS flow cytometer (Coulter Corp.). The histograms display relative cell numbers (*y* axis) as a function of relative fluorescence intensities (*x* axis).

Flow cytometry can also be used to check the specificity of Fab generated from IgG, instead of using indirect immunofluorescence. This involves the same protocol as with indirect immunofluorescence (*Protocol 3*) but instead of mounting the cells on a microscope slide, the cells are suspended in PBS for flow cytometry. Flow cytometry provides a much more objective result than indirect immunofluorescence (*Figure 4*), and is preferred where available.

5. Imaging with phycobiliproteins

Before the fluorescent probes can be used to image cells the fluorescent intensities and relative distributions of the individual probe molecules need to be established. This can be done by imaging the probes on poly-L-lysine slides. In this procedure microscope slides are coated with poly-L-lysine and probe is added directly onto the slide. The probe sticks to the poly-L-lysine and the individual probe molecules can then be imaged (*Figure 5*). From the distribution of the intensities observed it is then possible to determine what fluorescent intensity one would expect to see when imaging a single, double, triple, etc. fluorescent particle on the cell surface.

Protocol 7. Preparation of fluorescent probe poly-L-lysine slides

Equipment

- 70°C oven

Method

1. Immerse glass microscope slides[a] in alcoholic HCl (6% HCl, 70% EtOH) in a glass slide holder for 1 h.
2. Wash the slides nine times with ddH_2O.
3. Soak the slides in poly-L-lysine (0.01%) for 10 min at room temperature.
4. Rinse the slides once in ddH_2O and dry at 70°C for 25 min or overnight.
5. Prepare three different dilutions of the fluorescent probe and spread 15 μl over the centre of a slide. Leave for 10 min at room temperature.
6. Remove the probe with a Gilson pipette and wash twice, gently with 30 μl ddH_2O. Do not aspirate.
7. Add 15 μl ddH_2O to the slide, add a coverslip, and seal with silicone grease.

[a] Prepare five slides for each probe and choose the best three to use.

There are various ways to grow cells for fluorescence imaging, but ideally 8-well LabTek chambers (Nunc) should be used. These are expensive and the

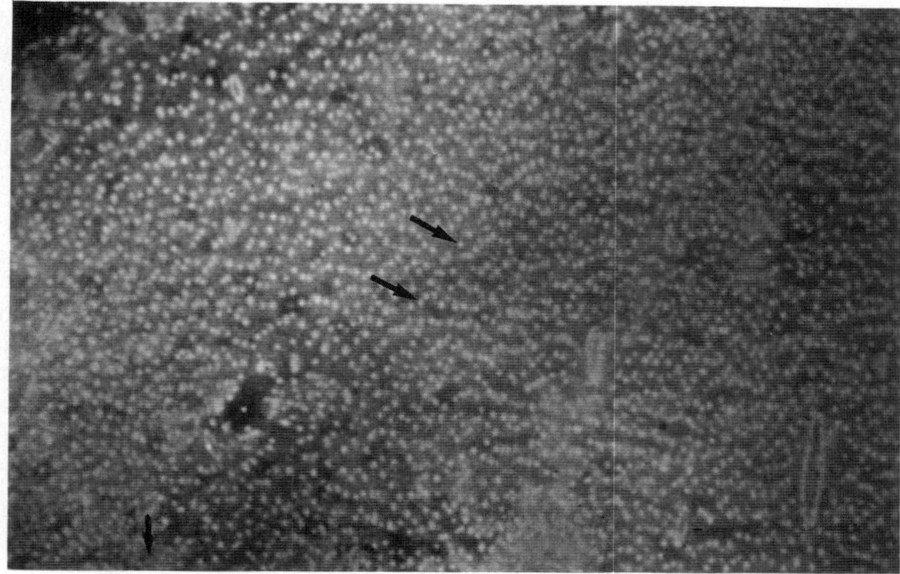

Figure 5. Digital flourescence image of single R–phycoerythrin labelled Fab particles (arrowed) bound to a poly-L-lysine coated microscope slide. Measuring the integrated fluoresence intensity of a single particle in this way allows dimers, trimers, and larger cluster sizes to be distinguished from single particles when the labelled Fab is bound to cells. This image was obtained using a cooled, slow-scan, CCD camera (Wright Instruments Ltd, Enfield, UK) controlled by Wright Instruments AT1 image control software, running on an IBM-AT compatible computer.

cells do not adhere as well to glass as to plastic, but they have the advantage that all the incubation of probe can be done in the individual 450 μl capacity wells, so using less probe. Also, the eight wells can be used for different dilutions of probe or even different probes. The culturing of cells and the incubation and washing conditions have to be ascertained for individual cell lines. In general we have found that two to three days after the cells have been added to the wells they are sufficiently adherent to withstand washing. The washing procedure needs to be very gentle making use of a pipette and not aspirating under vacuum.

The choice of cell line is particularly important. Aside from the obvious point of suitable cell surface antigens, the cells need to form monolayers and adhere well to the slide for individual fluorescent particles to be correctly imaged, for example fibroblasts are ideal. If the cells are not flat on the slide but have become slightly 'rounded', such as cells in suspension, only a limited area of the cell surface will be in focus making it difficult to identify individual fluorescent particles. *Figure 6* illustrates the type of images that can be obtained when cells are imaged using phycobiliproteins linked to Fab. Here, a

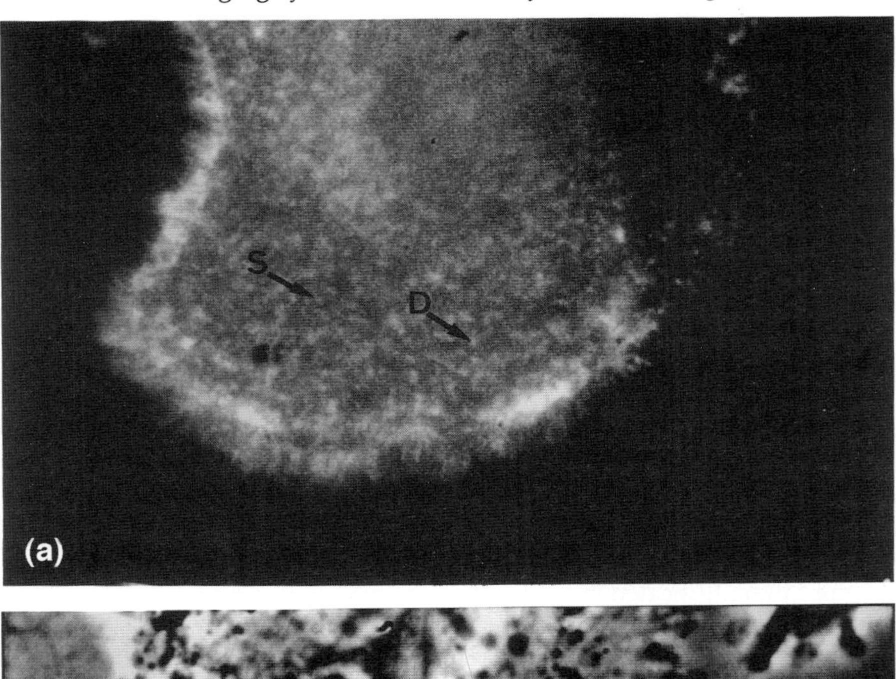

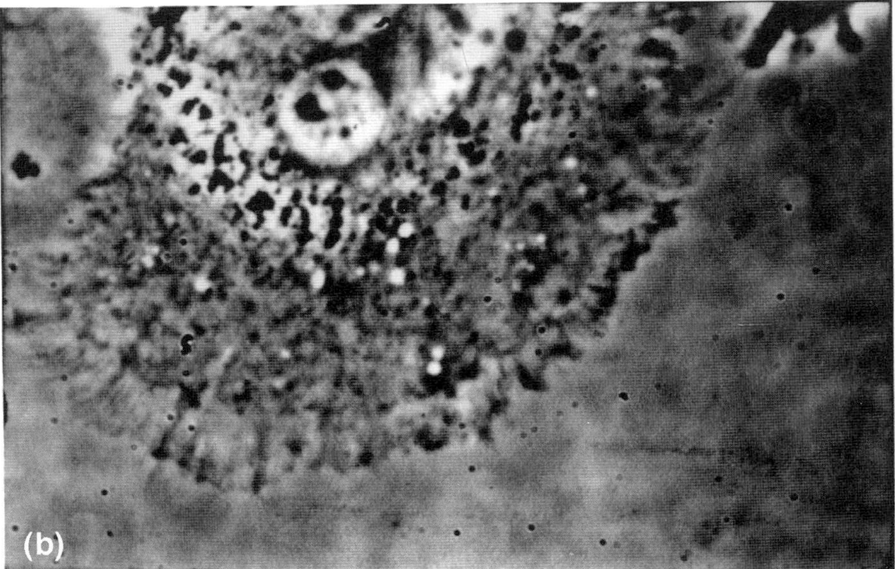

Figure 6. A digital fluorescence image (a) and phase-contrast image (b) of R--phycoerythrin labelled Fab bound to human class II DR molecules on mouse fibroblasts transfected with human HLA-DR alpha and beta genes. Single DR molecules (S) can be distinguished from dimers (D) by comparison of the relative fluorescence intensities. This image was obtained using a cooled, slow-scan, CCD camera (Wright Instruments Ltd, Enfield, UK) controlled by Wright Instruments AT1 image control software, running on an IBM-AT compatible computer.

Fab–R–phycoerythrin probe was prepared as described in this chapter, and used to image human class II DR1 molecules on the surface of mouse fibroblasts transfected with human HLA-DR alpha and beta genes. Analysis of images such as that shown in *Figure 6* enables the position and intensities of individual fluorescent spots to be determined. The intensities contain information on whether the spots consist of single particles or clusters of two or more. The positions of the particles are determined to within 25 nm and thus analysis of a sequence of images allows the movements of the receptor–particle complexes to be tracked with high spatial resolution.

Fluorescent imaging requires only the addition of a sensitive detection system to the video part of a standard fluorescence microscope. This detector could be an image-intensified video camera, or a cooled slow-scan, charge-coupled device (CCD). The first operates at video rate ($\approx$ 30 frames/sec) but needs frame averaging, while the latter will need exposures of 1 sec or more while the CCD integrates the fluorescence. A wide variety of such devices are available, operating usually from IBM-type personal computers, sometimes needing additional hardware such as image storage cards. Software to run such systems is also readily available, ranging from simple image acquisition/display programs to sophisticated image analysis and manipulation packages.

A 50 W mercury lamp is adequate for all except allophycocyanin which requires a laser. Single particle detection does require rather more attention to be given to the optical filters used in the fluorescence microscope. Standard filter blocks from microscope suppliers may give such high background signals that the phycobiliprotein fluorescence does not exceed the detector signal:noise limitation. Well designed filter assemblies are available for most common microscopes from, e.g. Omega Optical (USA), or via Glen Spectra, Stanmore Middlesex, UK. Further details of the specialized image analysis and tracking procedures can be found in refs 2 and 18.

We are currently using these procedures to track individual class I and class II MHC surface antigens on several cell types. The initial data indicate that some antigens undergo random diffusion whilst others are restricted to move within submicrometre scale domains. Yet others exhibit directed motion (Wilson *et al.*, in preparation) (23, 24, 25). The information obtained from these experiments will help to elucidate the cell surface organization of MHC molecules that is required for effective T cell recognition.

Acknowledgements

This work is supported by the UK Biotechnology and Biological Sciences Research Council (BBSRC) and by the University of Essex Research Promotion Fund.

References

1 Janeway, C. A. Jr. and Travers, P. (1994). *Immunobiology: the immune system in health and disease.* Blackwell Scientific Publications.
2. Morrison, I. E. G., Anderson, C. A., Georgiou, G. N., Stevenson, G. V. W., and Cherry, R. J. (1994). *Biophys. J.*, **67**, 1280.
3. Bormann, B. J. and Engelman, D. M. (1992). *Annu. Rev. Biophys. Biomol. Struc.*, **21**, 223.
4. Metzger, H. (1992). *J. Immunol.*, **149**, 1477.
5. Chakrabarti, A., Matko, J., Barisas, B. G., and Edidin, M. (1992). *Biochemistry*, **31**, 7182.
6. Capps, G. G., Robinson, B. E., Lewis, R. D., and Zuniga, M. C. (1993). *J. Immunol.*, **151**, 159.
7. Pawelec, G., Fernandez, N., Brocker, T., Schnieder, E. M., Festenstein, H., and Wernet, P. (1988). *J. Exp. Med.*, **157**, 243.
8. Fernandez, N., Hitman, G. A., Festenstein, H., Garde, C., Labeta, M. O., Walker-Smith, J. A., *et al.* (1988). *J. Clin. Exp. Immunol.*, **72**, 362.
9. Schick, M. R. and Levy, S. (1993). *J. Immunol.*, **151**, 4090.
10. Brown, J. H., Jardetzky, T. S., Gorga, J. C., Stern, L. J., Urban, R. G., Strominger, J. L., *et al.* (1993). *Nature*, **364**, 33.
11. Matko, J., Bushkin, Y., Wei, T., and Edidin, M. (1994). *J. Immunol.*, **152**, 3353.
12. Mecheri, S., Edidin, M., Mittler, R., and Hoffman, M. (1990). *J. Immunol.*, **144**, 1361.
13. Mecheri, S., Dannecker, G., Dennig, D., and Hoffman, M. (1990). *J. Immunol.*, **144**, 1369.
14. Jovin, T. and Vaz, W. L. C. (1989). In *Methods in enzymology* (eds. Fleischer, S. and Fleischer, B.), Vol. 172, pp. 471–573.
15. Peters, R. (1991). In *New techniques of optical microscopy and microspectroscopy* (ed. R. J. Cherry), pp. 199–228. Basingstoke: The Macmillan Press.
16. Barak, L. S. and Webb, W. W. (1981). *J. Cell. Biol.*, **90**, 595.
17. Barak, L. S. and Webb, W. W. (1982). *J. Cell Biol.*, **95**, 846.
18. Anderson, C. M., Georgiou, G. G., Morrison, I. E. G., Stevenson, G. V. W., and Cherry, R. J. (1992). *J. Cell Sci.*, **101**, 415.
19. Mage, M. G. (1980). In *Methods in enzymology* (eds Van Vunakis, H. and Langone, J. J.), Vol. 70, pp. 142–50.
20. Molecular Probes Catalogue. (1995).
21. Kurstak, E. (1986). *Enzyme immunodiagnosis.* Academic Press.
22. Ormerod, M. G. (ed.) (1990). *Flow cytometry: a practical approach.* IRL Press.
23. Smith, P. R., Wilson, K. M., Morrison, I. E. G., Koukidou, M., Fernandez, N., and Cherry, R. J. (1996). *Hum. Immunol.,* **47**, 96.
24. Wilson, K. M., Morrison, I. E. G., Smith, P. R., Fernandez, N., and Cherry, R. J. (1996). *J. Cell Sci.,* **109**, 2101–09.
25. Wilson, K. U., Morrison, I. E. G., Cherny, R. J., and Fernandez, N. (1997). In HLA: Biology and Medicine. XII HLA CONFERENCE. PARIS.

8

Sequence analysis of major histocompatibility molecules

DAVID A. LAWLOR and PURNIMA LAUD

1. Introduction

The GenBank database for nucleotide sequences contains over 1 500 000 entries and 1.5% are derived from loci of the major histocompatibility complex (MHC). Parham noted that in 1986 approximately 12 000 amino acid residues had been determined for human class I molecules by protein and DNA sequencing methods. He attributed the exponential growth of sequence information (fivefold increase in database size in less than three years) to the replacement of protein sequencing with the more efficient DNA sequencing procedures (1). At today's rate, the derived amino acid information for all MHC molecules increases by that amount every two weeks. *Table 1* lists the MHC sequence entries in GenBank for the various classes and species. The entries range in size from less than a 100 bp to the 66 kb fragment in the class II region recently sequenced by Beck and Trowsdale which includes *LMP2*, *TAP1*, *LMP7*, *TAP2*, and *DO*β (2). Although the class II region dominates the list with twice as many entries as the class I, the most striking feature of the list is increasing interest in sequencing MHC genes in species other than human or mouse.

This phenomenal sequencing effort reflects the tremendous genetic polymorphism which characterizes many of the major histocompatibility complex loci, such as *H2-K*, *D*, *Ab*, *Eb*, *HLA-B*, and *DRB1*, all having more than 60 alleles. It is well appreciated that the nucleotide substitutions distinguishing the MHC alleles generally introduce protein polymorphisms influencing interactions with peptide (3, 4). This consistent impact on protein function provides the major impetus for derivation of nucleotide sequences.

The diversified nature of the loci only invites repeated sequencing forays as they usually result in identification of novel alleles. And in light of the discoveries (5, 6) with isolated South American tribes of numerous HLA-B alleles not found in North American Indian populations, or other racial groups, it is clear that new alleles will be identified, especially in poorly characterized, reproductively isolated indigenous tribes, and the process of allelic generation

Table 1. GenBank/EMBL MHC submissions

Species	Class I	Class II	Others[a]
Human	230	590	31
Mouse	187	186	43
Fish	24	113	–
Gorilla	17	24	–
Rat	17	22	–
Monkey	16	74	–
Chimpanzee	13	153	–
Cow	11	36	–
Lizard	7	–	–
Hamster	6	–	–
Pig	6	27	2
Rabbit	5	7	–
Sheep	5	24	–
Chicken	5	5	10
Horse	3	29	–
Wallaby	3	7	–
Snake	3	–	–
Squirrel	2	13	–
Orangutan	2	10	–
Frog	2	8	–
Cat	1	–	–
Dog	1	18	–
Axolotl	1	–	–
Gibbon	–	8	–
Shark	–	2	–
Mole rat	–	3	–
Fox	–	1	–
Goat	–	1	–
Totals	567	1361	86

[a] Sequences from MHC loci other than class I or class II.

is ongoing at a palpable rate. The processes and forces which generate and select the allelic variants are still controversial (although less so now) and these considerations ensure that molecular evolutionists will continue their efforts to decipher patterns within MHC sequences.

2. Sequence analysis packages and operating systems

In this chapter, we want to acquaint readers with the various systems and programs which are available for genetic sequence analysis. It is not intended to be a comprehensive overview but rather to highlight programs employed in the authors' laboratory. The discussion will not concern the acquisition of the nucleotide sequence but rather concentrate on procedures that can be utilized after the sequencing project has been completed with all the compressions and ambiguities resolved. Although analysis can accompany a DNA sequencing project with useful information retrieved during the laboratory phases, invariably many analytical procedures will have to be repeated following determination of the complete, hopefully accurate, sequence. So it is largely left to the discretion of the researcher as to how far they wish to pursue analysis of unpolished sequence.

There are a large number of sequence analysis packages available to researchers. This chapter will focus on programs and packages which are available to the scientific community having been developed by programmers, population geneticists, and molecular biologists. Many of the programs (*PAUP*, *Phylip*, and *MEGA*) are offered at little, if any, cost to the community and represent a major dedication of effort by these developers.

2.1 Wisconsin sequence analysis package of the Genetics Computer Group

The Wisconsin Package was originally developed by John Devereux and Paul Haeberli in the laboratory of Oliver Smithies at the University of Wisconsin. Genetics Computer Group (GCG) remained associated with the university until 1990 when they formed a private company which derives its income from the fees charged to users. The licensing fee for the software is $4000 for academic sites or $12 000 for commercial companies. This fee includes maintenance, which entails any updates to the software within the first year, and technical support. For subsequent years the maintenance fees decrease to $3000 or $6000 for the two categories of users. The present version (9.0) of the Wisconsin Package is suitable for VAX/VMS, AXP/Open VMS, Sun Sparc, and Silicon Graphics Systems.

As an additional service, GCG provides seven databases which include GenBank, EMBL (modified), SwissProt, REBASE (restriction enzymes), and Transcription Factors. The databases are updated every other month and can be purchased individually or on an annual basis; current costs are $400/update or $2400/year for academic sites, and $600/update or $3600/year for commercial sites. The extensive documentation for the package, which is routinely updated, costs US$165 plus shipping. Inquiries can be directed to Dina Beers, Genetics Computer Group, Inc., University Research Park, 575 Science Drive, Madison, Wisconsin, USA 53711-1060 (e-mail: `DINA.Beers@GCG.com`, FAX: 608-231-5202).

The package contains over 100 programs which are grouped according to function such as sequence database searching, DNA fragment assembly, mapping, comparison, protein analysis, RNA secondary structure, and multiple sequence analysis (7). Each program does a distinct, well-defined function and generally accepts an input file and generates an output file. The output file often serves as the input file for the next program. Moving through the programs is aided by a number of prompts which helps set the parameters for the program. The default values, suggested by the program, are generally appropriate and provide a good starting point for the analysis. The user can change the parameters by typing the desired values at the prompt. Alternatively, the user can introduce the program modifiers on the command line and forego the process of sequential prompts. Usually there are fewer than six prompts per program so it requires little time to initiate a program regardless of approach. Some of the programs, such as the database searching programs, are computing-intensive and system managers will often request that these programs be initiated as a background or batch process. This redirection is easily done on the command line. There is also on-line help available.

The size of the package and its global approach, coupled with the computing power of mainframe computers, establish it as the optimal interface for sequence analysis. If the researcher only had access to one software package, the Wisconsin Package would be the hands-down choice. The few holes in the assortment of programs are filled by smaller packages with much more defined focus which are often designed for the personal computer.

2.2 Molecular Evolutionary Genetics Analysis (*MEGA*)

MEGA is a software package created by Sudhir Kumar, Koichiro Tamura, and Masatoshi Nei at the Institute of Molecular Evolutionary Genetics of Pennsylvania State University (8). The software is designed for IBM and IBM-compatible personal computers with hard disks, 640 KB RAM memory, and a MS-DOS version 3.3 or higher. Operation of *MEGA* does not require extended or expanded memory, or graphics adapters; a math co-processor is not required but will increase the speed and performance of *MEGA*. An extensive manual provides information regarding installation and required format for data files as well as guidelines in choosing various distance estimators. It also includes discussion of basic sequence statistics, tree-making procedures, the user-interface, and carefully detailed teaching examples using appropriate sample data files. Technical support, which is available for registered users via e-mail and fax, is quite prompt. Price of the package is $15 and covers the cost of the user manual, the disk, and mailing-handling expenses. Order forms can be obtained from the authors at the Institute of Molecular Evolutionary Genetics, 328 Mueller Laboratory, The Pennsylvania State University, University Park, PA, USA 16802 (e-mail: `imeg@psuvm.psu.edu`).

2.3 Phylogenetic Analysis Using Parsimony (*PAUP*)

PAUP is a powerful tree-making program written by David L. Swofford (9). The program is designed for the Macintosh and will run on almost all Apple computers and is fully compatible with system 7 and 68040 Quadra systems. The current version, 3.1.1, is accompanied by a 250 page instruction manual which contains a wealth of information regarding phylogenetic inference. The manual is divided into four chapters which cover background information (general concepts including character types, outgroups, search algorithms, consensus trees, tree-to-tree distances), program instructions, descriptions of the commands, and the user interface. The program, with on-line help access, utilizes standard Macintosh interface with menus, dialogue boxes, and scrollable lists. A number of sample files are provided which can be employed to gain familiarity with the numerous options available. There is also editor function associated with the program enabling correction of file formatting errors. *PAUP* provides technical support by e-mail, standard mail, fax, and express courier, although careful reading of the manual will resolve most problems. It is also routinely updated by an 'updater' program which corrects any 'bugs' users may have spotted during their sessions. The program markets for $100 and ordering information is available from PAUP Support, Laboratory of Molecular Systematics, MRC 534, MSC, Smithsonian Institution, Washington, D.C. 20560 USA (Internet: `paup@onyx.si.edu`).

2.4 Phylogeny inference package (*Phylip*)

Phylip is a free package of programs (version 3.5) designed by Joe Felsenstein of the University of Washington for inferring phylogenies. It consists of 31 programs for analysis of molecular sequence data and distance matrix data. Also included are programs for plotting cladograms, phenograms, and consensus trees. The package is distributed on diskettes for use on Macintosh and IBM personal computers; network distribution using a file transfer protocol is another option and details can be obtained by e-mail (`joe@genetics.washington.edu`). As the programs are written in a standard computer language, C, they can also be run on UNIX and VAX/VMS systems.

2.5 Commercial packages for personal computers

Other packages are commercially available such as Microgenie from Intelligenetics, LaserGene from DNASTAR, or MacDNASIS from Hitachi Software. The commercial programs are designed for the personal computer and are reported to be quite all-inclusive, providing DNA sequencing and analytical capability. However, they are not inexpensive and other drawbacks include periodic updates of databases and working in a computing environment with less than mainframe capability. These may be the best option for some users working in environments without access to the Wisconsin Package, or for those desiring the freedom of an independent system not subject to system shutdowns or CPU charges.

3. Database search programs

Generally, the first step in analysing a newly obtained sequence is to compare it to the entries contained in the various databases (GenBank, EMBL,

SwissProt). This will not only indicate whether the sequence is novel but also identify homologous sequences that reside in the databases. The Wisconsin Package has a number of search programs that can be employed for this preliminary analysis. *FASTA*, *TFASTA*, *WORDSEARCH*, and the *BLAST* programs will perform pairwise comparisons of the query sequence against all of the database sequences and generate an output file with information on the most closely related sequences. *FASTA* and *TFASTA* were created by Pearson and Lipman (10), *WORDSEARCH* uses an algorithm developed by Wilbur and Lipman (11), and the *BLAST* series employs methods of Karlin and Altschul (12).

3.1 Query sequences and sample data file

In order to highlight the capabilities of a number of programs, we utilized a number of sequences as input files (*Figure 1*). The sequences were chosen as there was considerable information concerning locus and function.

The first query sequence is Gogo-Oko (`ggoko.gb_pr`), a gorilla class I MHC molecule with chimeric composition of an MHC-A background with a 210 bp component from MHC-H. The H-component occurs in exon 2 which encodes the α1 domain of the class I heavy chain; the rest of the molecule

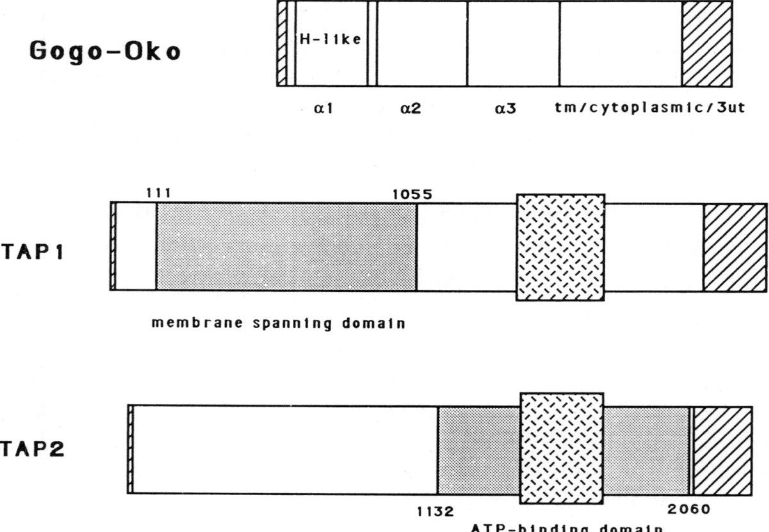

Figure 1. Schematic of the three molecules used as input sequences for the analytical programs. The class I MHC sequence Gogo-Oko (1267 bp) contains the entire coding sequence and portion of the 3′-untranslated region. The 210 bp H-like component encodes much of the α1 domain. The other two test sequences are a 945 bp fragment from the variable 5′ half of the human TAP1 cDNA and a 928 bp fragment from the conserved 3′ half (containin the ATP-binding motif) of the gorilla TAP2.

resembles MHC-A. This chimeric gene is found in appreciable frequency in the gorilla and is thought to have resulted from a gene conversion event between the *H* and *A* loci. It was chosen for inclusion to test the ability of the database search programs to discern the 'H' segment in an otherwise normal looking MHC-A molecule. The query sequence (1267 bp) contains the entire coding and portion of 3′ untranslated region.

The other two query sequences are 1 kb fragments from the human *TAP1* and gorilla *TAP2* genes, genes which have been mapped to the class II region in the MHC and encode proteins involved in the transport of peptides to the site of class I heavy chain synthesis in the endoplasmic reticulum. Both sequence similarity and chromosomal placement strongly suggest that a gene duplication event generated the pair of TAP genes. The TAP1 cDNA (`hsring4.gb_pr`) fragment is the 5′ half and it encodes the multiple membrane spanning domains postulated to form the pore for peptide transport. It is the more variable portion of *TAP1* (and *TAP2*) and it was included to test the sensitivity of the database searches to detect homologous sequences. The gorilla TAP2 query sequence is a 1 kb fragment from the 3′ half of the coding region and it contains the sequence motif responsible for binding ATP and which groups the molecules in the ABC-family of transport proteins. This fragment of the molecule is well conserved and was included to determine whether the database searches could effectively segregate similar sequences into the categories of orthologous, paralogous, or distantly related genes.

The sample data file for the tree-making procedures consists of the coding region (1098 bp) for 49 MHC-A locus alleles from hominoid species (human, common chimpanzee, gorilla, orangutan, and gibbon), and *HLA-B2705* as an outlier for the data set. The data set was compiled to test the ability of the tree-making programs to handle large number of sequences.

3.2 *FASTA*

FASTA searches can be performed with either nucleotide or peptide query sequences and the output file consists of three parts, a histogram derived from the search, a list of the highest ranking database entries, and the pairwise comparisons of the region of greatest similarity for each of the high scoring sequences. The query sequence is fragmented into words which comprise a dictionary; the default wordsizes for nucleotides and peptides are 6-mers and 2-mers, respectively. As the search progresses through the database, the individual entries are broken into words and these are compared to the words of the query sequence. If identities are found, they are listed as hash marks on a graph with one axis consisting of the query sequence and the other axis, the database sequence. The ten regions of greatest similarity are noted and the single highest score from this group is designated init1. A rescoring is performed with less stringent conditions thereby allowing conservative substitutions or shorter stretches of identity to contribute to the secondary score grade. Also, at this stage, *FASTA* attempts to join some of the initial regions of similarity and the subsequent score is referred to as initn. The initn score is used to rank the database sequences and the histogram presents the distribution of sequences with their scores. The final (optimized) score results from a re-inspection of the init1-generated alignments for further relatedness between the query and library entry (13).

Figure 2 shows partial output from a *FASTA* search of the combined GenBank and EMBL (GenEMBL) databases with the gorilla TAP2 fragment. In this example, the score column represents the initn score and this ranges from < 4 to > 160. The histogram is compressed so that each line represents a bin of four initn scores, 1–4, 5–8, 9–12, and so forth. The highest initn score listed is > 160 and there are 44 sequences which achieved this rank. Of the 44 sequences, 42 of them exceeded the 160 with the initial scoring grade, init1. The remaining two sequences attained the > 160 only after the secondary grading, initn. Although the optimization procedure to obtain the initn generally increases the score of the database sequence, scores can go down occasionally. There were 49 sequences which grouped with scores of 81–84 after the primary scoring, init1. After rescoring to produce initn, only 37 of the original 49 maintained that rank.

The symbols used to form the histogram reflect whether the init1 or initn scores predominate the group. Equal (=) marks represent the sequences for which both the init1 and initn attained the score level; plus (+) symbols represent sequences which only attained the scoring level with the initn, and minus (−) symbols are for sequences with init1, but not initn, scores of that level. Following the histogram are the mean init1 and initn scores for all the sequences in the database.

The second part of the output file is the listing of the best scores against the gorilla TAP2 fragment. Only 16 of the 50 listings are shown in the figure but they reveal four categories of similar sequences. The first four sequences are from orthologous genes in other species, human and rat. The TAP2E and PSF-2 mRNA, independent accessions to the database, correspond to human *TAP2*; the third highest initn score is with the GenBank entry `hsmhcapg.gb_pr` which is a 66 kb DNA contig that contains *TAP2*, as well as *TAP1*, *LMP2*, *LMP7*, and *DOB*. The marked variations in the init1, initn, and optimization scores are due to comparison of a cDNA query sequence with a very large stretch of genomic DNA that includes multiple loci. The rat TAP2 sequence is the fourth member of this group and its scores are comparable to the first three.

The second group of four consist of sequences derived from the closely related gene, *TAP1*, in human, rat, and mouse. The initn scores for this group range from 443–902. Although `hsring4` and `hsy3` are human TAP1 sequences, there are marked differences in scores due to the extent of comparison with the gorilla TAP2 query sequence (900 bp with `hsring4` and 400 bp for `hsy3`). The sequence similarity and genomic arrangement of *TAP1*

```
(Nucleotide) FASTA of: okotap2frg  from: 1 to: 927  October 3, 1993  22:38

  REVERSE-COMPLEMENT of: 2oa1101g2.frg  check: 4736  from: 1  to: 927

 TO: GenEMBL:*  Sequences:    155,863  Symbols: 173,636,677  Word Size: 6

Score Init1 Initn
<  4   820   820:=================================================
   8     0     0:
  12     2     2:=
  16     1     1:=
  20    35    35:==================
  24  4477  4477:=================================================
  28 19506 19506:=================================================
---------------------- // ---------------------------- // -----------------
  84    49    37:===================------
  88    14    14:=======
  92     3    18:==+++++++
  96     4    88:==++++++++++++++++++++++++++++++++++++++++++
 100     3    99:==++++++++++++++++++++++++++++++++++++++++++++++++
 104     1    78:=++++++++++++++++++++++++++++++++++++++
 108     4    48:==++++++++++++++++++++++
 112     2    27:=+++++++++++++
 116     1    13:=++++++
 120     1    10:=++++
 124     0     7:++++
 128     1     1:=
 132     1     4:=+
 136     2     4:=+
 140     0     4:++
 144     0     2:+
 148     1     3:=+
 152     0     0:
 156     0     0:
 160     0     0:
>160    42    44:=====================+
 mean initn score:  36.4 (5.45)
 mean init1 score:  36.4 (5.45)


The best scores are:                                           init1 initn opt..

gb_pr:hstap2ea  Z22936 H.sapiens TAP2E mRNA, complete CDS...2654  2654  3622
gb_pr:hummhpsf2a  M74447 Human PSF-2 mRNA, complete cds. ...2647  2647  3608
gb_pr:hsmhcapg  X66401 H.sapiens genes TAP1, TAP2, LMP2, ... 766  2397   773
gb_ro:rnmtp2  X63854 Rat mRNA for transporter polypeptide...1676  2299  2337
gb_pr:hsring4  X57522 H.sapiens RING4 cDNA 7/92              697   902  1128
gb_ro:rnmtp1  X57523 R.norvegicus mtp1 mRNA 6/92             719   778  1116
gb_ro:musmhcham  M55637 Mus musculus HAM1 gene, complete ... 571   684  1047
gb_pr:hsy3  X57521 H.sapiens partial mRNA Y3 for peptide ... 443   443   566
gb_ro:musmdra  J03398 Mouse mdr gene encoding a multidrug... 224   389   526
gb_in:dromdr65  M59077 Drosophila melanogaster P-glycopro... 163   206   535
gb_pr:hsmdr3  X06181 Human mRNA 3'-fragment for P-glycopr... 129   199   189
gb_ov:papgpb3  X72068 P.americanus gene for P-glycoprotei... 189   189   213
gb_ro:rnplect  X59601 Rat mRNA for plectin 8/91               66   146    66
gb_ba:pprpocg /rev  X16538 P.putida rpoC gene for the bet...  69   145    71
gb_ba:rhmndva  M20726 R.meliloti ndvA gene encoding a 67.... 106   143   374
gb_pr:hsjhcmu /rev  X56795 Human IgM heavy chain switch r...  64   140    64
```

Figure 2. Partial output file from a *FASTA* search with the gorilla TAP2 nucleotide sequence. The output consists of a histogram of similarity scores, followed by a list of sequences found by the search, and finally an alignment (not shown) of the query sequence with each of the listed sequences. Though the *R. meliloti ndvA* gene sequence (gb_ba:rhmndva) is listed towards the end, it exhibited a significant homology with gorilla TAP2 (34% identity over 320 amino acids).

and *TAP2* argue that the genes resulted from a duplication event and the *FASTA* search supports a paralogous relationship.

As the TAP2 query sequence contains the ATP-binding site that defines the ABC transporter family, it is not surprising that the third group of sequences consist of members of that family. The mouse multidrug resistance and various P-glycoprotein genes have slightly lower initn scores (189–389) than the TAP1 comparisons.

Although the fourth group of sequences have initn scores (140–146) which are significantly higher than the mean, their homology to the gorilla TAP2 sequence was questionable. The only one of the four which is clearly related to TAP2 is the *ndvA* gene from *Rhizobium meliloti*. The pairwise comparisons, which comprise the third portion of *FASTA* output (data not shown), revealed a 53% identity over 440 bp. Although the extended length of comparison suggested relatedness, it required translating the ndvA nucleotide sequence and comparing the peptide with the TAP2 protein. This comparison revealed 34% identity over 320 amino acids. The *R. meliloti ndvA* gene encodes a glucan transporter which contains motifs characteristic for ATP-binding transporters. Analysis of the other three database sequences in this final group did not reveal any common features that could be attributed to ancestry. The limited stretches of comparison are probably random occurrences with no evolutionary significance. The ndvA sequence is informative since there is a noticeable increase in *FASTA* score (143–374) with optimization. This spike pattern can help guide one in discriminating related sequences from mere random matches. The other three sequences (plectin, rpoC, and human IgM heavy chain switch region) show decreases in scores upon optimization.

3.3 *TFASTA*

TFASTA is a variation of *FASTA* that requires a peptide query sequence. Each nucleotide database entry is translated in the six reading frames (top and bottom strand) and the query is compared against the individual translations. A search method similar to that used for *FASTA* is employed and the resultant output file is composed of the same three parts: histogram, the list of best sequences with init1, initn, and optimized scores, and the pairwise comparisons with the query peptide. *Figure 3* shows the ninth of the 40 pairwise comparisons contained in the *TFASTA* output file with the human TAP1 fragment (`ring4.pep`) as query sequence. This sequence, consisting of 315 residues, is aligned with the human TAP2 sequence, `hstap2ea.gb_pr` which received the accession number z22936 at the time of submission to GenBank/EMBL. The search program detected similarity with the query peptide and the second (2) reading frame of `hstap2ea`. Also, listed are the init1 (246), initn (280), and optimized (313) scores.

This comparison was chosen for display since it highlights a rule of thumb for discriminating randomly matched sequences with those that are ancestrally

```
ring4.pep
gb_pr:hstap2ea

LOCUS         HSTAP2EA      2522 bp     RNA      PRI          17-SEP-1993
DEFINITION    H.sapiens TAP2E mRNA, complete CDS.
ACCESSION     Z22936
KEYWORDS      ABC transporter gene; TAP2E.
SOURCE        human
  ORGANISM    Homo sapiens . . .

SCORES        Frame: (2) Init1: 246 Initn: 280 Opt: 313
               26.9% identity in 286 aa overlap

               10        20        30        40        50        60
ring4. TALPRIFSLLVPTALPLLRVWAVGLSRWAVLWLGACGVLRATVGSKSENAGAQGWLAALK
                                     |||:  |:||  :|:  :    :| |: :
hstap2 PWTSLLLVDAALLWLLQGPLGTLLPQGLPGLWLE--GTLR--LGGLWGLLKLRGLLGFVG
             40        50        60          70          80

               70        80        90       100       110       120
ring4. PLAAALGLALPGLALFRELISWGAPGSADSTRLLHWGSHPTAFVVSYAAALPAAALWHKL
       :|  :| || |  : :|:|:: :::::::::   :|:     ::|:|:||  : :||  |
hstap2 TLLLPLCLATPLTVSLRALVAGASRAPPARVASAPWSW----LLVGYGAAGLSWSLWAVL
       90       100       110       120           130       140

                 130       140       150       160       170
ring4. ---GSLWVPGGQGGSGNPVRRLLGCLGSETRRLSLFLVLVVLSSLGEMAIPFFTGRLTDW
          |:     :| ::   ::|||    :: : |   : ::||: |||  || ::||::|
hstap2 SPPGAQEKEQDQVNNKVLMWRLLKLSRPDLPLLVAAFFFLVLAVLGETLIPHYSGRVIDI
           150       160       170       180       190       200

       180       190       200       210       220       230
ring4. ILQDGSADTFTRNLTLMSILTIASAVLEFVGDGIYNNTMGHVHSHLQGEVFGAVLRQETE
       :  | ::::|:::: :|::::::|:: :   :| :: ||:::: ::::::|:::|||: :
hstap2 LGGDFDPHAFASAIFFMCLFSFGSSLSAGCRGGCFTYTMSRINLRIREQLFSSLLRQDLG
           210       220       230       240       250       260

       240       250       260       270       280       290
ring4. FFQQNQTGNIMSRVTEDTSTLSDSLSENLSLFLWYLVRGLCLLGIMLWGSVSLTMVTLIT
       |||:::||:: ||:::||: :|: |: | :::|: ||: : | |:||  | :||:::|:
hstap2 FFQETKTGELNSRLSSDTTLMSNWLPLNANVLLRSLVKVVGLYGFMLSISPRLTLLSLLH
           270       280       290       300       310       320

       300       310
ring4. LPLLFLLPKKVGKWYQLL
       :|: :   |  ::::|
hstap2 MPFTIAAEKVYNTRHQEVLREIQDAVARAGQVVREAVGGLQTVRSFGAEEHEVCRYKEAL
           330       340       350       360       370       380
```

Figure 3. Partial output file from a *TFASTA* search showing an alignment of the query sequence (human TAP1 peptide) with the human TAP2 sequence. Query sequence must be in protein format. *TFASTA* searches nucleotide database (GenEMBL) and converts nucleotide sequences into protein (all six reading frames). The *TFASTA* output format is identical to the *FASTA* output—listing a histogram of similarity scores, followed by a list of sequences found, and finally, alignments between the query sequence and each of the listed sequences.

related. Doolittle suggests that if two sequences are > 25% identical over a 100 residue stretch it is reasonable to assume that they diverged, relatively recently, from a common ancestor (14). In this case, the human TAP1 and TAP2 are 26.9% identical in a 286 amino acid overlap. This degree of identity is even more striking considering that the variable, amino termini are the basis for the comparison. Even though this portion of the molecule has evolved at a faster rate than the carboxyl terminus, it is still possible to discern the evolutionary relatedness.

3.4 *WORDSEARCH* and *BLAST*

Other programs serve functions comparable to *FASTA* and *TFASTA*. *WORDSEARCH* and the *BLAST* series of programs are available with the Wisconsin Package and the quality of the searches with the same query sequences appear comparable to that obtained with *FASTA* and *TFASTA*. The *BLAST* series can input nucleotide or peptide sequences for search of the respective databases and *TBLASTN* is comparable to *TFASTA* with a protein query sequence being compared to a nucleotide sequence database dynamically translated in all reading frames. The *BLAST* series is a relatively recent addition to the package and claims are made for greater speed of database searches than with the *FASTA* series. A faster search, if it didn't adversely affect sensitivity, would be preferable, especially if CPU costs are a major consideration. All of the search programs are computer-intensive (if applied to a large database such as GenBank/EMBL) so system managers generally request that they be performed as a background (batch) operation. This helps reduce the workload on the system and saves the user money as batch CPU is generally charged at a lower rate than interactive CPU. As a test for significant differences in search speeds with the programs, the query sequences were inputted to the programs using the default search parameters. In all three cases, *WORDSEARCH* was considerably slower than the other two programs (105–113 versus 48–53 CPU minutes). The search times with the three query sequences were reduced (5–10%) with *BLASTN* in two of the three instances. The savings in CPU time, with *FASTA* and *BLASTN* searches, is relatively insignificant unless a large number of database searches are conducted.

The more important question concerning database searches is the sensitivity. Although the programs appeared to perform equally well in recovering the expected sequences from the database, there were instances of recoveries with some of the programs but not others. For instance, the *FASTA* search with Gogo-Oko ranked several HLA-H sequences within the top 15 scores; *WORDSEARCH* ranked HLA-H within the top 30; *BLASTN* had over 200 sequences with higher scores than HLA-H in its output, and *TFASTA* didn't turn up the homologous sequence within the top 50 scores. Gogo-Oko was included as a query sequence due to the presence of the 'H'-component and its quick detection by *FASTA* speaks well for the search program.

However, when the TAP1 fragment was the query sequence, *TFASTA* and *WORDSEARCH* appeared to be more sensitive than *FASTA*. All the programs listed the TAP1 homologues (murine, rodent, and human) as best scores but only *TFASTA* and *WORDSEARCH* showed the homologies in the TAP2 database sequences. *FASTA* and *BLASTN* were insensitive in this specific instance and didn't list TAP2 sequences in their best scores. When TAP2 fragment was the query sequence, no major differences were observed in output files for the four programs. The list of best sequences were largely overlapping and no obvious oversights were made. All of the search programs found the TAP2 homologues, the TAP1 sequences and multiple members, both eukaryotic and prokaryotic, of the ABC-binding transporter family. It is possible to increase sensitivity of the searches by decreasing stringency of parameters, such as wordsize, but, as noted by Pearson, this is often accompanied by decreasing selectivity (13). Numerous authors strongly suggest utilizing peptide comparisons in searches for distantly related sequences as there is greater discriminatory value with 20 amino acids than four bases and it eliminates the problems of codon usage bias and degeneracy (13–15). Therefore, the preferable programs would be *TFASTA* and *TBLASTN* which compare peptide query sequences against the six possible reading frames of the entries in the large DNA databases. Our searches of protein databases (SwissProt) with a peptide query has not been particularly revealing which we attribute to differences in database size. The protein databases are becoming secondary depots for molecular entries and should not be relied on exclusively for obtaining homologous sequences; for instance, the only TAP-related sequence obtained from SwissProt search with the gorilla TAP2 fragment was the mouse HAM1.

4. Multiple sequence alignment

The database searches will locate the sequences, if any, that are homologous to the sequence of interest. These homologous sequences can be imported into one's directory with the program *FETCH* of the Wisconsin Package. The sequences acquired in such fashion will be a hodge-podge of genomic and cDNA sequences of various lengths. At this time it is generally necessary to conform the query and homologous sequences to ensure they are colinear before pursuing more intensive analysis.

4.1 *PILEUP*

A recent addition to the Wisconsin Package, the multiple sequence alignment program, *PILEUP*, can effect this function. The program can align up to 300 sequences of a maximum length of 5000 by a series of pairwise alignments. The program requires as an input file, a file-of-file names (FOF). If the database search revealed 25 sequences clearly homologous with the query sequence, these would be imported with *FETCH* and a file could be created using an on-line editor such as emacs or vi. The contents of this file consists solely of the names of the related sequence files and, of course, the sequence of interest. This FOF (after being named `ex5pep.fof` in this example) would then be the appropriate response at the prompt:

Pileup of what sequences? `@ex5pep.fof`

The ampersand always precedes the fof and alerts the program to be ready for a file-of-file names and not an individual sequence. Of course for the program to run, the individual sequence files must reside within the same directory. An example of output from *PILEUP* is shown in *Figure 4.* The default parameters for gap weight (3.0) and gap length weight (0.1) were accepted; however, for complicated alignments, it is best to experiment with these values and compare the resulting output files. Fortunately, class I gene

```
PileUp of: @ex5pep.fof

 Symbol comparison table: GenRunData:pileuppep.cmp  CompCheck:
1254

                  GapWeight: 3.0
            GapLengthWeight: 0.1

 ex5pep.msf  MSF: 39  Type: P  October 22, 1993  15:12  Check:
807  ..

 Name: HLA-A0201       Len:   39  Check: 9682  Weight:  1.00
 Name: HLA-A1101       Len:   39  Check: 9792  Weight:  1.00
 Name: HLA-Cw0101      Len:   39  Check:  369  Weight:  1.00
 Name: HLA-Cw0201      Len:   39  Check:  369  Weight:  1.00
 Name: Gogo-C0201      Len:   39  Check:  714  Weight:  1.00
 Name: HLA-B2705       Len:   39  Check: 9572  Weight:  1.00
 Name: HLA-B4001       Len:   39  Check: 9572  Weight:  1.00
 Name: Gogo-B0101      Len:   39  Check: 9919  Weight:  1.00
 Name: Gogo-C0101      Len:   39  Check: 9870  Weight:  1.00
 Name: Gogo-B0201      Len:   39  Check:  948  Weight:  1.00


            1                                         39
 HLA-A0201  PSSQPTIPIV GIIAGLVLFG A.VITGAVVA AVMWRRKSS
 HLA-A1101  LSSQPTIPIV GIIAGLVLLG A.VITGAVVA AVMWRRKSS
HLA-Cw0101  PSSQPTIPIV GIVAGLAVLA VLAVLGAVVA VVMCRRKSS
HLA-Cw0201  PSSQPTIPIV GIVAGLAVLA VLAVLGAVVA VVMCRRKSS
Gogo-C0201  PSSQPTIPIV GIVVGLAVLV VLAVLGAVVT AMMCRRKSS
 HLA-B2705  PSSQSTVPIV GIVAGLAVLA V.VVIGAVVA AVMCRRKSS
 HLA-B4001  PSSQSTVPIV GIVAGLAVLA V.VVIGAVVA AVMCRRKSS
Gogo-B0101  PSSQSTIPIV GIVAGLAVLA V.VVIGAVVT AVICRRKSS
Gogo-C0101  PSSQPTIPIV GIVAGLAVLA V.VFTGTVVA AVMCRRKSS
Gogo-B0201  PSSQSTIPIV GIVAGLAVLV VTVAVVAVVA AVMCRRKSS
```

Figure 4. Alignment of a set of hominoid class I MHC alleles using *PILEUP*. Sequences were grouped in a file using an on-line editor (emacs). A single gap was inserted in some of the sequences by *PILEUP* to produce an optimal alignment. Default parameters were accepted. All of the input sequences were given an equal weight by the program.

products show little size polymorphism and only minor alignments are needed in the transmembrane domain (exon 5) to ensure colinearity. This pileup consists of the membrane spanning domain for products from the *A*, *B*, and *C* loci. The program inserted a single gap to produce alignment. This was an easy task and could have been performed by hand. However, more difficult alignments, which are required for comparisons of class I and II, or TAP1 and TAP2 molecules, are best performed with the aid of a computer program. Optimal alignment of multiple sequences is a difficult and computationally-intensive procedure; this is best evidenced by the relative scarcity of multiple sequence alignment programs. *PILEUP* determines the two most similar sequences in the fof and aligns them to produce a cluster. In progressive fashion, it brings in the next most similar sequence and accommodates it in the alignment with insertion of gaps. Before the program begins the process of aligning the sequences with gap insertions, it produces a dendrogram of the clusters and this tree can be subsequently loaded into the program *FIGURE* for graphics output.

5. Tree-making procedures

One of the most useful measures employed to analyse sequences is tree-making. The tree can readily reveal the sequences that are most similar and can usually assign sequences to loci. Although there are pitfalls associated with inferences of evolution (especially with MHC genes which bear remarkable signs of frequent intralocus and occasional interlocus genetic exchange), the tree's graphic clustering of sequences conveys relationships very effectively. There are numerous methods to produce gene trees and all the packages mentioned earlier include tree-making programs. In fact for *PAUP*, *MEGA*, and *Phylip*, tree-making is a prominent, if not sole, feature of the package. The Wisconsin Package has recently added the capability of producing UPGMA (unweighted pair group method with arithmetic mean) trees as an associated output from the sequence alignment program, *PILEUP*.

5.1 *PAUP*

PAUP utilizes the principle of parsimony to produce trees from nucleotide or peptide sequences. An excellent synopsis of parsimony for inferring phylogeny can be found in the review by Stewart (16).

There are three algorithms which can be employed depending on the size of the input file. The heuristic search can output trees with large numbers of operational taxonomic units (OTUs). In the sample data file, there are 50 OTUs (sequences) with 1098 bp and the authors have generated *PAUP* trees with 80–90 OTUs of comparable length and are unaware of upper limits imposed by the program. The heuristic method does not ensure that the optimal tree(s) (most parsimonious) will be found but requires far less computer time than the exact methods. The more stringent algorithms do have limitations concerning the number of OTUs they can handle. The branch-and-bound algorithm can accommodate 12–15 OTUs, but computing time becomes excessive with greater number, and probably shouldn't be considered an option if OTUs approach 18. The exhaustive search will evaluate every possible tree that can be generated from the data set and produce the optimal tree(s) but the required computing time restricts this form of tree-making to data sets of 10–11 OTUs. Since *PAUP* can be run in background using the multifinder, the user doesn't have to dedicate the Macintosh for tree searches but can utilize the computer for other purposes while the search proceeds.

Generally, large data sets of MHC alleles will result in numerous 'most parsimonious' trees and *PAUP* can produce consensus trees from this set. The *PAUP* tree in *Figure 5* is a 'strict' consensus tree of the homonoid A locus alleles. The heuristic search with the data set resulted in 1471 equally parsimonious trees after evaluating over 49×10^6 rearrangements. The computing time required for the search was nearly 22 hours on a Macintosh IIsi. The 'strict' tree is the most conservative estimate of consensus and *PAUP* provides the user the capability to generate semistrict, majority-rule, and Adams consensus trees which preserve more defined structure than the strict. If the heuristic method with the large data set presents the user with a multiplicity of 'most parsimonious' trees, the regions of ambiguity can often be resolved by employing the exhaustive methods to a subset of the OTUs.

For instance in this tree a number of ambiguities occur in the groupings of the HLA-A2 subtypes and the A19 cross-reactive group of antigens (HLA-A2901, -A3101, -A3201, -A3301). The A2 family could be selected for analysis by branch-and-bound method and the A19 family, with seven members, could be approached with the exhaustive method. After resolution of a particular group, the set of 'most parsimonious' heuristic trees could be perused and the tree with the desired topology for that group of sequences could be selected. An alternative to this dissection of a large tree into discrete component branches is to delete OTUs from the data set. This can often be done without adversely affecting the interpretation of the gene tree. For instance, it is not necessary to include all the molecular subtypes for the HLA-A2 antigen in assessing relatedness of the hominoid A locus alleles. Reducing the OTUs to a number of suitable for analysis with the stringent algorithms provides the user confidence that the shortest tree represents the 'true' tree. An inherent danger with heuristic algorithm is it may be confined within a cul de sac of locally optimal trees and fail to uncover the globally optimal tree.

5.2 *MEGA* tree-making algorithms

The *MEGA* program enables users to produce *UPGMA*, neighbour-joining (*NJ*), and maximum parsimony trees. For maximum parsimony trees, two forms of search are possible: the branch-and-bound and heuristic. For the *NJ* and *UPGMA* trees, statistical evaluation is possible with the bootstrap test.

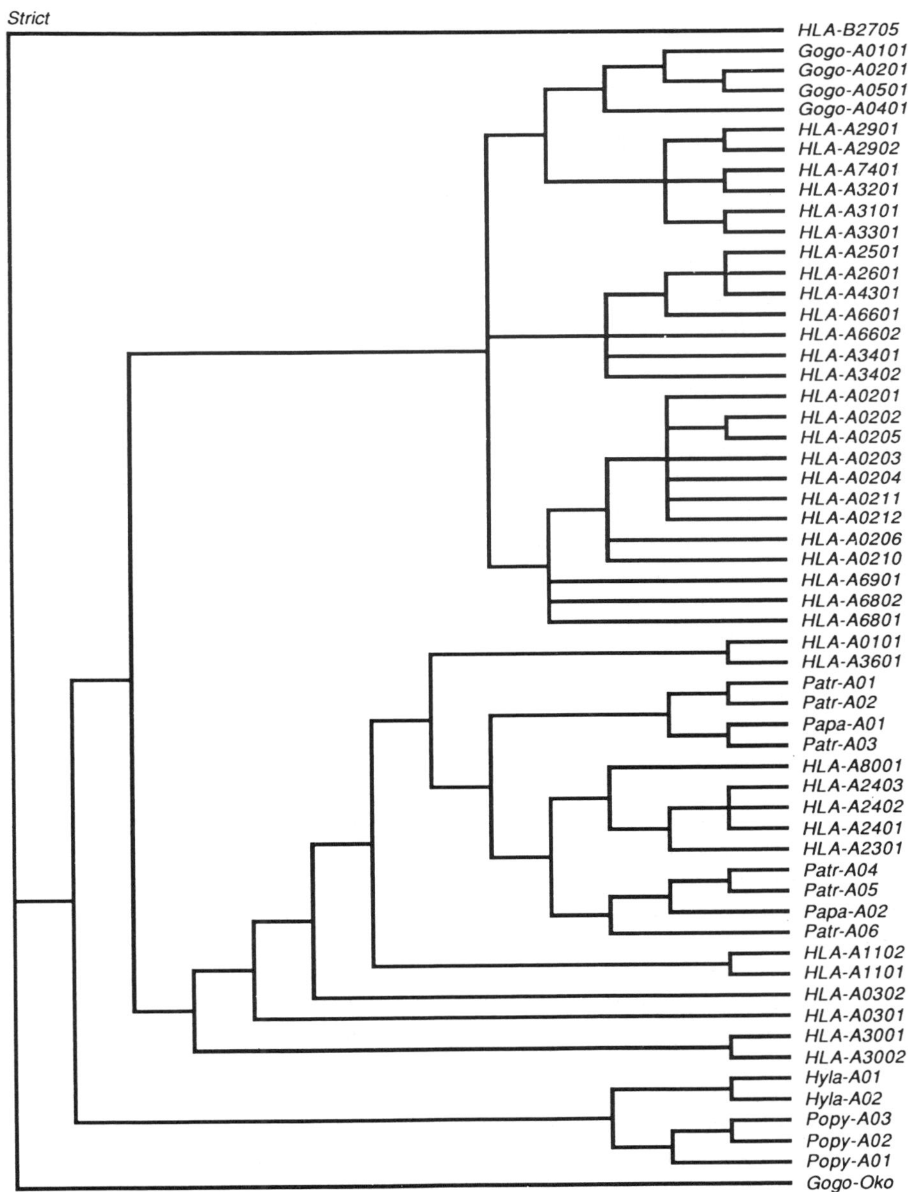

Figure 5. Strict consensus *PAUP* tree of the hominoid MHC-A locus alleles. A general heuristic search with simple addition of OTUs and tree bisection-reconnection swapping algorithm produced 1471 'equally parsimonious' trees with a length of 656. The data file consisted of 50 OTUs with 1098 characters (> 190 being parsimony informative characters). HLA-B2705 served as the outlier to root the tree.

The package also provides the phylogenetic tree editor which enables the user to swap descendent branches, display the mirror image of the tree, and root the tree on any chosen branch. The neighbour-joining method searches for pairs of OTUs that effectively reduce branch lengths. The pair is referred to as neighbours since they are connected by a single interior node. This pair would then be regarded as a single 'OTU' (OTU 1 + 2) and be paired with the next taxonomic unit (OTU 3) that most successfully minimizes total length of branches. There are N − 2 pairs of neighbours obtainable from N OTUs and, with successive addition of neighbouring pairs, the tree is constructed (17). The *NJ* method, differing from parsimony methods, calculates nucleotide or amino acid differences to create a distance matrix that is used to determine neighbour pairs.

The *NJ* tree in *Figure 6*, obtained from the same data set as the *PAUP* tree, required 20 minutes of computing time on a 386-IBM compatible machine. Gogo-Oko is the most divergent of the sequences, other than the outlier, HLA-B2705,; this deep branch for Gogo-Oko is due to the 'H'-like fragment found in this gorilla class I molecule. In comparing the *NJ* and *PAUP* trees, there are no major discrepancies in placement of the sequences to branches. The authors have generally noted concordance with *NJ* and *PAUP* trees of the same data set. Although the methods of computing the trees are considerably different with *NJ* employing distance measurements and *PAUP* using parsimony informative positions, the trees generally provide comparable interpretations to the data set.

6. Calculation of nucleotide substitution rates

A powerful assessment of the impact of positive selection on the diversification of MHC genes is provided by calculating nucleotide substitution rates at synonymous and non-synonymous sites. Hughes and Nei were the first to utilize the statistical method to provide evidence for positive selection of nucleotide substitutions in codons comprising the antigen binding site (3). Synonymous substitutions do not result in amino acid replacements and therefore these mutations are independent of Darwinian positive selection which acts at the protein level. Non-synonymous substitutions do result in coding changes and the variant protein, if benefiting from the mutation, would be positively selected. Alternatively, the amino acid replacement could impair the function of the protein and purifying selection would attempt to eliminate the protein and silence the allele. As a general rule of thumb, synonymous substitution rates should exceed non-synonymous rates and the instances where this doesn't hold true are regarded as possible examples for positive selection.

The *MEGA* program calculates nucleotide distances, both total and synonymous/non-synonymous, as well as peptide distances. Variations on the distance estimations, available as sub-menu options, include Jukes-Cantor,

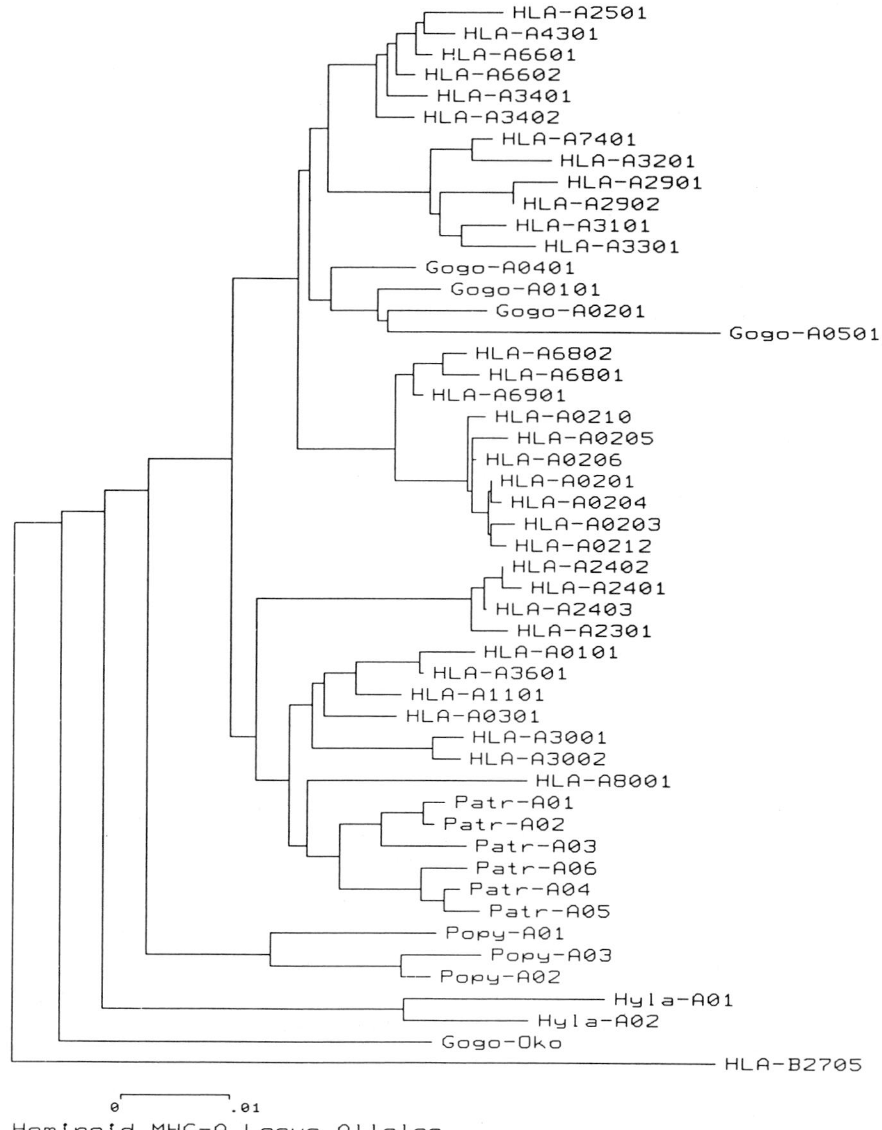

Figure 6. Neighbour-joining tree of the hominoid MHC-A locus alleles. The data set was the same as used in *Figure 5*. The distance measure employed was the number of nucleotide differences with the Jukes-Cantor correction. Branch lengths are measured in the number of nucleotide substitutions per site.

Tajima-Nei, Tamura, Tamura-Nei, and Kimura two-parameter distances. Factors which would influence one's choice for distance estimations are sequence size, nucleotide frequencies, and transition/transversion biases, and the user's manual discusses these considerations and makes recommendations for the method best suited for the user's data set. For instance, the Jukes-Cantor correction for distance estimations, often employed for nuclear genes, provides accurate values for nucleotide substitutions if the base frequencies and transition/transversion ratios are approximately equivalent. Kimura's two-parameter distance estimation is recommended for analysing data sets, such as mitochondrial DNA, which have higher rates of transitional than transversional nucleotide substitutions.

In *Table 2*, *MEGA* calculations for synonymous (d_s) and non-synonymous (d_n) substitution rates for four groups of MHC-A locus alleles are shown. Analysis of the five HLA-A sequences in `humhla.meg` (one of several example data files which comes with the package) shows that non-synonymous substitution rate exceeds the synonymous substitution rate in the 57 codons that encode the antigen binding site but not in the other two regions. This pattern of increased rate of non-synonymous substitutions within the antigen binding site is also seen with the five gorilla (Gogo), six common chimpanzee (Patr), and the large panel of 49 hominoid-A locus alleles. However, the comparison reveals that the Patr-A alleles have a considerably lower rate of non-synonymous substitutions at this functional site than the other groups (7.2 vs. $>$ 13.7). Therefore the effects of positive selection, although evident, are not as pronounced within the A locus of the common chimpanzee.

7. Motif analysis

A recent addition to the Wisconsin Package is the program, *MOTIFS*, which searches for amino acid patterns found in the Prosite Dictionary of Protein Sites and Patterns. The dictionary, compiled by Dr Amos Bairoch of the University of Geneva, contains short sequence patterns occurring in proteins. Although many of the entries are post-translational sites for glycosylation or phosphorylation, more specific patterns such as leucine zippers are also contained in the dictionary. The search algorithm will not tolerate gaps in the query peptide but ambiguities can be compensated for with the command line instruction, `–MISmatch=n`. Partial outputs from the program with Gogo-Oko and TAP2 peptides as input sequences are shown in *Figures 7* and *8*.

The only motif revealed, for Gogo-Oko, is the imunoglobulin and major histocompatibility complex protein signature which is centred at the second cysteine involved in the intradomain disulfide bond of the α3 domain of the class I heavy chain. This default search was stringent as mismatches were not allowed and the sequence pattern found in Gogo-Oko corresponds exactly to the dictionary septamer motif of (F, Y) x C x (V, A) x H, where the first residue has to be either F or Y, followed by any residue, then C, any residue, V

Table 2. Synonymous (d_s) and non-synonymous (d_n) substitution rates in different regions of MHC-A locus alleles

	Antigen binding site (57 codons)[a]		**Other α1, α2 positions (125 codons)**		**α3 Domain (92 codons)**	
	d_s	**d_n**	**d_s**	**d_n**	**d_s**	**d_n**
HLA-A (5) [b]	6.2 ± 2.6[c]	13.7 ± 2.3	2.9 ± 1.3	1.8 ± 0.5	9.4 ± 3.0	1.5 ± 0.6
Gogo-A (5)	5.4 ± 2.6	14.8 ± 2.3	8.2 ± 2.1	3.5 ± 0.7	2.8 ± 1.5	2.3 ± 0.7
Patr-A (6)	0.7 ± 1.0	7.2 ± 1.6	2.7 ± 1.3	0.3 ± 0.2	2.8 ± 1.6	0.6 ± 0.4
Hominoid-A (49)[d]	4.8 ± 1.6	14.5 ± 1.4	5.8 ± 1.1	2.1 ± 0.4	7.9 ± 2.0	2.1 ± 0.6

[a] 57 codons in the α1 and α2 domains encoding positions that comprise the antigen binding site.
[b] Five human A locus alleles from the `humhla.meg` file in *MEGA*.
[c] Jukes-Cantor estimate of mean nucleotide substitutions and standard errors presented as percentages.
[d] 49 MHC-A locus alleles from human (HLA), gorilla (Gogo), common chimpanzee (Patr), gibbon, and orangutan.

```
MOTIFS from: Gogo-Oko.pep

Mismatches: 0                  October 13, 1993  13:25  ..

Gogo-Oko.pep  Check: 6934  Length: 366   ! TRANSLATE of: gogo-oko
check: 8818 from: 1 to: 1098
________________________________________________________________

IG_MHC                    (F,Y)xCx(V,A)xH
                            (Y)xCx(V)xH
            281: GKEQR        YTCHVQH       EGLPK

*****************************************************************
* Immunoglobulins and major histocompatibility complex proteins
signature *
*****************************************************************

It is known [4,5] that the Ig constant chain domains and a single
extracellular  domain  in  each  type  of  MHC  chains  are  related.
These homologous domains are approximately one hundred amino acids
long, and they include a conserved intradomain disulfide bond.  We
developed a small pattern around the C-terminal cysteine involved
in this disulfide bond which can be used to detect these category
of Ig related proteins.

-Consensus pattern: [FY]-x-C-x-[VA]-x-H
________________________________________________________________
```

Figure 7. Partial output file from the *MOTIFS* program. Gorila class I molecule (peptide) was used as query sequence. *MOTIFS* identified an immunoglobulin signature (intradomain disulfide bond) with an exact match. No degeneracy was introduced into the search (mismatch = 0). Five amino acids on either side of the motif are also listed. Examples of proteins with the same motif and relevant citations are also included in the program output (not shown).

or A, any residue and, finally, H. This pattern occurs in Gogo-Oko at positions 286–292 and the flanking residues are also presented. This signature sequence is found in other members of the Ig supergene family such as β2-microglobulin and class II MHC molecules. *MOTIFS* can also reveal common patterns found in many proteins as the result of post-translational modifications. The program will assess the input peptide for *N*-glycosylation sites, protein kinase C and tyrosine kinase phosphorylation sites, *n*-myristoylation, and amidation sites. As these motifs are largely uninformative regarding protein relationships, the option is suppressed and requires the command line modifier `-FREquent`. The program, surprisingly, did not reveal other motifs for the class I molecule. The dictionary doesn't presently contain motifs for the CD8 binding site that is well conserved in the hominoids or the Bw4 and Bw6 motifs which divide B locus molecules in humans, apes, and monkeys (18, 19). The program's dependence on a short linear arrangement of residues for motif assignment precludes it from being able to detect complex motifs which exist spatially, such as the conserved tyrosines (Y7, Y59, Y159, Y171) which contribute to the A pocket in the antigen binding site of class I molecules.

The search with the TAP1 peptide revealed no matches which was not surprising since this portion of the molecule is variable and lacks known

```
MOTIFS from: okotap2frg.pep

Mismatches: 1              October 28, 1993  15:25  ..

     okotap2frg.pep  Check: 5840  Length: 309    ! TRANSLATE of:
okotap2frg check: 4050 from: 1 to: 927
```

```
ATP_GTP_A                 (A,G)x4GK(S,T)
                           (G)x{4}GK(S)
           135: VTALV    GPNGSGKS    TVAAL

****************************************
* ATP/GTP-binding site motif A (P-loop) *
****************************************

From sequence comparisons and crystallographic data analysis it has been
shown [1,2,3,4,5] that an appreciable proportion of proteins that bind
ATP or GTP share a number of more or less conserved sequence motifs.  The
best conserved of these motifs is a glycine-rich region, which probably
forms a flexible loop between a beta-strand and an alpha-helix.
There are numerous ATP- or GTP-binding proteins in which the P-loop is
found.  We list below a number of protein families for which the
relevance of the presence of such motif has been noted:

  - ATP synthase alpha and beta subunits.
  - Myosin heavy chains.
  - Nitrogenase iron protein family (nifH/frxC).
  - ATP-binding proteins involved in 'active transport' [6].
  - GTP-binding elongation factors (EF-Tu, EF-1alpha, EF-G, EF-2, etc.).

-Consensus pattern: [AG]-x(4)-G-K-[ST]

[ 1] Walker J.E., Saraste M., Runswick M.J., Gay N.J. EMBO J. 1:945-
951(1982).
[ 2] Moller W., Amons R. FEBS Lett. 186:1-7(1985).
[ 3] Fry D.C., Kuby S.A., Mildvan A.S. Proc. Natl. Acad. Sci. U.S.A.
83:907-911(1986).
```

Figure 8. Partial *MOTIFS* output using gorilla TAP2 fragment as query peptide. Degeneracy was introduced into the search by allowing one mismatch (using qualifier -MISmatch=1 on the command line) and post-translational modifications of the protein were listed in addition using the qualifier -FREquent. The above motif detected by the search is an exact match. However, other motifs with mismatches were also revealed, such as tyrosine kinase phosphorylation sites, amidation sites, etc.

functional motifs. In contrast, the TAP2 peptide search turned up the ATP/GTP-binding site motif A (P loop) which characterizes the ABC family of transporters, as well as any number of other families of nucleotide-binding proteins (*Figure 8*). The output from *MOTIFS* lists the consensus pattern, other proteins which share the particular motif with the query peptide, and citations of relevant publications.

If the *MOTIFS* search is not as revealing as one expected, introduce degeneracy to the search. For example, the default TAP2 search (no mismatches) revealed the P loop motif but didn't uncover the signature motif for ATP-binding active transport proteins which is commonly found in the carboxyl terminus of these proteins. Initiating a search with the proviso that two mismatches should be considered does uncover the signature sequence (*Figure 9*).

The gorilla TAP2 sequence does have two mismatches with the consensus signature sequence with alanines at P2 and P3 instead of serine at P2 and serine/glycine at P3. At the other ten positions, TAP2 agrees with the

```
MOTIFS from: okotap2frg.pep

 Mismatches: 2                    November 3, 1993  10:03  ..

      okotap2frg.pep  Check: 5840  Length: 309    ! TRANSLATE of:
okotap2frg check: 4050 from: 1 to: 927
___________________________________________________________________

ATP_BIND_TRANSPORT
(L,I,V,M,F,Y)S(S,G)Gx3(R,K,A)(L,I,V,M,Y,A)x(L,I,V,M,F)(A,G)
                          (L)S(S)Gx{3}(R)(L)x(I)(A)
             239: EKGSQ         laagqkqrlaia                RALVR
mis=2

***********************************************************
* ATP-binding proteins `active transport' family signature *
***********************************************************

On the basis of sequence similarities a family of related ATP-
binding
proteins has been characterized [1 to 5].  These proteins are
associated with a variety of distinct biological processes in both
prokaryotes and eukaryotes, but a majority of them are involved in
active  transport of small hydrophilic molecules across the
cytoplasmic membrane.

In prokaryotes:

 - Active transport systems components: arabinose (araG); vitamin
B12 (btuD), molybdenum (modC); sulfate (cysA); iron(III) ...

 - Hemolysin/leukotoxin export proteins hlyB, cyaB and lktB.
 - Beta-(1,2)-glucan export proteins chvA and ndvA.
 - Polysialic acid transport protein kpsT [8].

In eukaryotes:

 - The multidrug  transporters (Mdr)  (P-glycoprotein), a family
of closely related proteins which extrude a wide variety of drugs
out of the cell (for a review see [11]).
 - Cystic fibrosis transmembrane conductance regulator (CFTR),
which is most probably involved in the transport of chloride ions.
 - Histocompatibility antigen modifier 1 (HAM1), which may be
involved in the transport of antigens from the cytoplasm to a
membrane-bound compartment for association with MHC class I
molecules [12].

As a signature pattern for this class of proteins, we use a
conserved  region which is located between the A and the B motifs
of the ATP-binding site.

-Consensus: [LIVMFY]-S-[SG]-G-x(3)-[RKA]-[LIVMYA]-x-[LIVMF]-[AG]
___________________________________________________________________
```

Figure 9. Partial *MOTIFS* output from a search using gorilla TAP2 fragment as query peptide. A mismatch of 2 was allowed in the search. Two alanines are present in the motif instead of a serine followed by either serine or glycine. This motif was not brought up by a search using a mismatch of 0 or 1.

signature motif. In all of the programs, the advantages of decreasing stringency to increase sensitivity is oftentimes accompanied by increasing the background of random selections. In the case of *MOTIFS*, it is warranted to decrease the stringency since the motifs themselves contain a high degree of ambiguity with multiple residues being permitted at certain positions.

7.1 *PROFILESCAN*

A complementary program in the Wisconsin Package is *PROFILESCAN* which uses the method of Mike Gribskov to find structural and sequence motifs in protein sequences (20). The representative motifs are contained as profiles in a library and the program will align each profile to the input sequence and display all alignments between the profile and sequence that exceed the set threshold score. The program also totals the number of alignments found with the profiles so it indicates presence of repeats or duplicated structures, such as a zinc-finger motif.

Figure 10 shows portions of the two output files from *PROFILESCAN* with Gogo-Oko. In *Figure 10a* there is comparison of the α3 domain of Gogo-Oko with the IgC-like profile. This is basically the same region detected by *MOTIFS* but is considerably larger than contained in the motifs output and includes the N terminal cysteine contributing to the disulfide bond in that domain. *Figure 10b* lists the number of occurrences for each motif in the sequence of interest. *PROFILESCAN*, in contrast to *MOTIFS*, did not locate the ATP/GTP-binding site in the TAP2 fragment suggesting deficiencies in the profile library. As the library increases with well-documented profiles, it will become only more valuable for detecting functional components of novel molecules.

8. Conclusion

It is evident that the keen interest in molecular characterization of MHC genes will not diminish in the foreseeable future. Whereas the earlier analysis focused on human and mouse molecules, recently, there has been a remarkable diversification of effort to examine many vertebrate species. Although the characterization of the human class I and II genes will continue until the genetic polymorphism has been completely described and the mechanisms responsible for locus diversification are understood, this shift in interest to other species will pick up pace. Also, there will be increasing interest in MHC genes which have been largely neglected or only recently identified. The genetic map of the human MHC contains over 100 genes and many of these will attract the interest of scientists intrigued with the evolutionary history of the complex (21). These recently identified genes may have function markedly different from the peptide-binding molecules which have dominated earlier studies and the first insights to their roles could well come from comparisons

(a)

```
PROFILESCAN of : Gogo-Oko.pep  check: 6934  from: 1  to: 366
_____________________________________________________________________

 Profile: ProfileDir:ig-constant.prf
   Gap weight:  3.00      Gap Length weight:   0.10
   Ave match:   0.14      Ave mismatch      :  -0.14
(Peptide) PROFILEMAKE v4.00 of: @CONSTANT.FIL   from: 1   to: 116
Length: 116
  Gap: 1.000000  Len: 1.000000
Sequences: 8  MaxScore: 43.85  11-AUG-1987 21:21
                         hlmstr.frg  From: 1     To: 1   Weight:1.00
                         rwhugc.frg  From: 1     To: 116 Weight:1.00
                      huigm-ch4.frg  From: 1     To: 116 Weight:1.00

Profile: ig-constant.prf     alignment: 1

 Quality: 16.98           Gaps: 3
 Ratio:   0.16             Length: 116
 Normalized quality: 2.22
                        .         .         .         .         .
S     199 GKETLQRTDPPKTHMTHHPVSDHEATLRCWALGFYPAEITL..TWQRDGE 246
          |..: .::    ::  .:. ....:||| |:| ||:|::|||   :|:::|.
P       1 GSVSAPSVFLLPPSSEQNINLSKSATLVCLVTDFFPADITVTISRKRDGS 50
                        .         .         .         .         .
S     247 DQTQDTELVETRPGGD....GTFQKWAAVVVPS...GKEQRYTCHVQHEG 289
          . :.:.. .:|....:    :||.: :.|.|:.   :|:::|||.|.|||
P      51 SLISGKTETSTSEADSNSGPGTYSASSTLTVSEEDRNSEERYTCQVTHEG 100
                        .
S     290 LPKPLTLRWEPSSQPT 305
          ||.|:| :  ::|  .
P     101 LPSPVTEKTVSPPSSS 116
_____________________________________________________________________
```

(b)

```
PROFILESCAN of : Gogo-Oko.pep  check: 6934  from: 1  to: 366

Compare to profile library: GenRunData:profilescan.fil

                   globin.prf       0 alignments over threshold
              ig-constant.prf       1 alignments over threshold
              ig-variable.prf       0 alignments over threshold
                    zinci.prf       0 alignments over threshold
                   zincii.prf       0 alignments over threshold
                ca-efhand.prf       0 alignments over threshold
                    hth21.prf       0 alignments over threshold
                   cyclic.prf       0 alignments over threshold
                    nuc10.prf       0 alignments over threshold
                    homeo.prf       0 alignments over threshold
                   kinase.prf       0 alignments over threshold
                     chap.prf       0 alignments over threshold
                     perf.prf       0 alignments over threshold
                     p450.prf       0 alignments over threshold
                    ssrna.prf       0 alignments over threshold
                   serine.prf       0 alignments over threshold
                      myc.prf       0 alignments over threshold
                      pou.prf       0 alignments over threshold
```

Figure 10. (a) First part of a *PROFILESCAN* program output. Gorilla class I MHC peptide (Gogo-Oko) was used in this search. An alignment between the query peptide and a sequence from the immunoglobulin-constant library is shown. The profile detected was also found by the *MOTIFS* search (see *Figure 7*) and is the immunoglobulin protein signature. (b) Second part of the *PROFILESCANE* program output. The number of alignments found in each of the profile libraries is listed. One alignment was found for the Ig-constant profile (alignment in *figure 10a*).

with databases. The experience with the antigen-processing genes in the class II region confirms the value of this approach in gaining early, rapid information with novel molecules (22–27).

Acknowledgements

The authors' research has been supported by a grant from NCI (CA16672).

References

1 Parham, P. (1988). *Am. J. Med.*, **85**, 2.
2. Beck, S., Kelly, A., Radley, E., Khurshid, F., Alderton, R. P., and Trowsdale, J. (1992). *J. Mol. Biol.*, **228**, 433.
3. Hughes, A. L. and Nei, M. (1988). *Nature*, **335**, 167.
4. Hughes, A. L. and Nei, M. (1989). *Proc. Natl. Acad. Sci. USA*, **86**, 958.
5. Belich, M. P., Madrigal, J. A., Hildebrand, W. H., Zemmour, J., Williams, R. C., Luz, R., *et al.* (1992). *Nature*, **357**, 326.
6. Watkins, D. I., McAdam, S. N., Liu, X., Strang, C. R., Milford, E. L., Levine, C. G., *et al.* (1992). *Nature*, **357**, 329.
7. Genetics Computer Group. (1992). *Program manual for the GCG package*, version 7. 575 Science Dr., Madison, Wisconsin USA 53711.
8. Kumar, S., Tamura, K., and Nei, M. (1993). *MEGA: Molecular Evolutionary Genetics Analysis*, version 1.0. The Pennsylvania State University, University Park, Pennsylvania 16802.
9. Swofford, D. L. (1993). *PAUP: Phylogenetic Analysis Using Parsimony*, version 3.1 Computer program distributed by the Illinois Natural History Survey. Champaign, Illinois.
10. Pearson, W. R. and Lipman, D. J. (1988). *Proc. Natl. Acad. Sci. USA*, **85**, 2444.
11. Wilbur, W. J. and Lipman, D. J. (1983). *Proc. Natl. Acad. Sci. USA*, **80**, 726.
12. Karlin, S. and Altschul, S. F. (1990). *Proc. Natl. Acad. Sci. USA*, **87**, 2264.
13. Pearson, W. R. (1990). In *Methods in enzymology*, Vol. 183, p. 63–98.
14. Doolittle, R. F. (1990). In *Methods in enzymology*, Vol. 183, p. 99–110. Academic Press.
15. Henikoff, S., Wallace, J. C., and Brown, J. P. (1990). In *Methods in enzymology*, Vol. 183, p. 111–132.
16. Stewart, C.-B. (1993). *Nature*, **361**, 603.
17. Saitou, N. and Nei, M. (1987). *Mol. Biol. Evol.*, **4**, 406.
18. Salter, R. D., Benjamin, R. J., Wesley, P. K., Buxton, S. E., Garrett, T. P. J., Clayberger, C., *et al.* (1990). *Nature*, **344**, 41.
19. Wan, A. M., Ennis, P., Parham, P., and Holmes, N. (1986). *J. Immunol.*, **137**, 3671.
20. Gribskov, M., Homyak, M., Edenfield, J., and Eisenberg, D. (1988). *CABIOS*, **4**, 61.
21. Campbell, R. D. and Trowsdale, J. (1993). *Immunol. Today*, **14**, 349.
22. Deverson, E. V., Gow, I. R., Coadwell, W. J., Monaco, J. J., Butcher, G. W., and Howard, J. C. (1990). *Nature*, **348**, 738.

23. Monaco, J. J., Cho, S., and Attaya, M. (1990). *Science*, **250**, 1723.
24. Spies, T., Bresnahan, M., Bahram, S., Arnold, D., Blanck, G., Mellins, E., *et al.* (1990). *Nature*, **348**, 744.
25. Trowsdale, J., Hanson, I., Mockridge, I., Beck, S., Townsend, A., and Kelly, A. (1990). *Nature*, **348**, 741.
26. Browne, M. G., Driscoll, J., and Monaco, J. J. (1991). *Nature*, **353**, 355.
27. Glynne, R., Powis, S. H., Beck, S., Kelly, A., Kerr, L.-A., and Trowsdale, J. (1991). *Nature*, **353**, 357.

9

Approaches to studying MHC expression in mammals other than primates

SHIRLEY A. ELLIS

1. Introduction

Until recently, nearly all studies on MHC expression and function had been carried out in humans and mice. In these species an enormous body of data has now been collected concerning the minute details of polymorphism present in the classical MHC class I and II genes, as techniques have progressed from serology, through biochemistry, to various molecular approaches. Using this information it has become possible to understand how MHC, peptide, and T cell receptor interact (1), to study the mechanisms involved in evolution of the MHC (2), and to speculate on the role of pathogens as a selective force in maintaining the high level of diversity seen in these molecules (3).

Interest in the MHC has widened in recent years to encompass studies on other primate species (4), and on very much more distantly related species such as fish, reptiles, and amphibians (5). However, there have been relatively few detailed studies of MHC in other mammals, despite the potentially useful nature of such information. Studies on the MHC of domestic animals such as cows, pigs, and sheep are essential for increasing our understanding of protective immune responses to many important pathogens which have the potential to cause enormous economic loss throughout the world (6, 7).

This chapter will describe the various approaches which can be used to design studies of MHC expression in a mammalian species (primarily for MHC class I, although similar approaches could be used to study class II), using the cow as an example.

2. Strategy

It is important when deciding on a strategy, to consider carefully what the information generated is to be used for. For example:

(a) Population analysis, i.e. how many alleles are present and what is their relative frequency.

(b) To establish which haplotypes/alleles are common, and then to do a detailed study on these.

(c) To aid with specific disease studies—to choose alleles associated with protection/susceptibility, or those which are very common in exposed populations.

It is essential to be aware of any relevant data which may be available on the species chosen. For example, are any serological reagents, or monoclonal antibodies to MHC available? Have any MHC sequences been published? There are three major options available (it may be desirable to use all three in combination, but this is not always possible):

- serological analysis
- biochemical analysis
- molecular analysis

These will be dealt with in turn in the next section. The example that will be used is the cow. The strategy that was chosen is outlined in *Figure 1*.

The aims of this particular study were to initially identify common haplotypes in a given population (the available experimental cow herd). The chosen haplotypes were then to be investigated in more detail. In order to do this it was necessary to establish how many class I genes were being expressed as part of the haplotype. As much detail as possible was to be generated about each allele. Disease studies were run in parallel using animals carrying the chosen haplotypes, and information and reagents (e.g. transfectants, monoclonal antibodies) could then be fed into these programmes as appropriate.

3. Methods

This study was initiated in 1990, and so the strategy was based on data available at that time. Four international workshops had been carried out to analyse bovine MHC (BoLA), and most of the information generated (class I) concerned serological specificities, of which there appeared to be approximately 30 (8). At this time there was little evidence to demonstrate the existence of more than one expressed class I locus. There were several monoclonal antibodies in existence which recognized a monomorphic determinant on bovine class I, e.g. ILA19 (9), also some raised against human MHC class I cross-reacted on bovine, e.g. W6/32 (10). This meant that it was possible to attempt some biochemical analysis of the expressed molecules.

It is worth noting that even if no species-specific anti-MHC reagents are available, it can be useful to check a panel raised against other species, as there is often cross-reactivity. However, in this case care must be taken, as some-

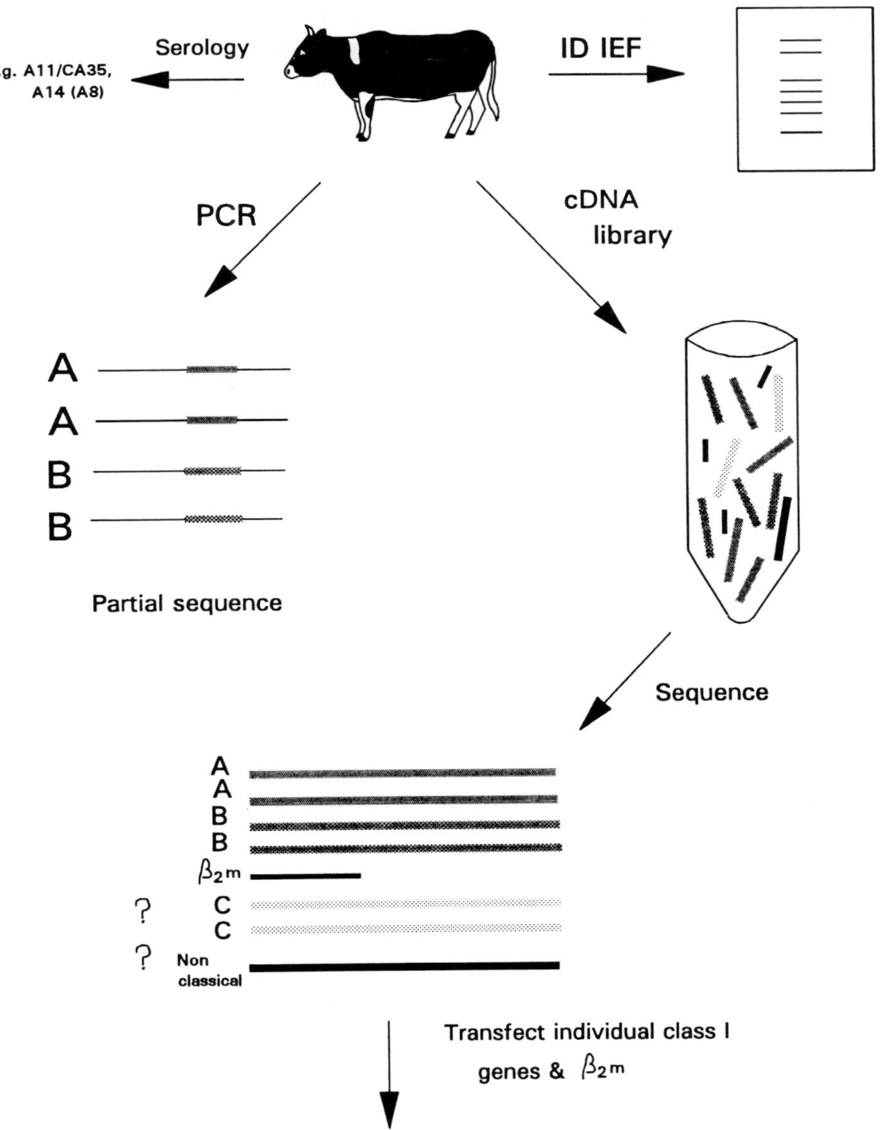

Figure 1. A strategy for detailed analysis of MHC class I haplotypes, demonstrating the information generated from three approaches – serology, 1D IEF and molecular.

times not all MHC products will be recognized. For example, although W6/32 will immunoprecipitate most bovine class I heavy chains, it does not react with a proportion (possibly the products of a single locus). If in doubt, therefore, it is worth using more than one antibody for detailed analysis.

Five bovine MHC class I cDNA sequences had been published at the beginning of the study, but only two of these had been related to serological specificities (11).

3.1 Serological analysis

Alloantisera are obtained from animals following pregnancy, or after specific immunization by skin grafting or lymphocyte inoculation. These are used in a standard complement-mediated lymphocyte microcytotoxicity assay (12). It is unclear in the case of many cattle typing sera whether they recognize a single allele, or a haplotype. Recent analysis of bovine class I transfectants has shown that in some haplotypes the serological specificity is carried on one gene product, with additional gene products being unrecognized (and serologically 'blank') (13). Such serological analysis is best done in a laboratory where it is a matter of routine. If such facilities are not available, or if sera have not already been reasonably well characterized, for example in a workshop as described, then it is best to avoid this step, as it is unlikely to generate meaningful data.

In our study we obtained typing data on a representative number of the population, and chose three common haplotypes as our starting point: A14, A11, and A18.

3.2 Biochemical analysis

One-dimensional isoelectric focusing (1D IEF) is a useful technique for studying polymorphism of MHC molecules. The method involves immunoprecipitation of radiolabelled heavy chains, followed by separation on an IEF gel, i.e. according to isoelectric point (pI). Detailed methodology is given in ref. 14. Although this technique is very sensitive and can separate variants which differ by only a few amino acids, some alleles may have an identical overall charge, and will not therefore be distinguished by this technique.

Some analysis of cattle MHC using this method has been carried out (15). Whereas in humans it has been possible to assign a defined A or B allele to each band observed (16), this has not proved to be the case with the cow, partly because of the complex banding patterns sometimes obtained (and because serology is not as informative). These data suggest that some MHC class I genes may be giving rise to more than one product, possibly as a result of alternative splicing, or post-translational modifications, as is observed in HLA-C (17). An alternative explanation, supported by recently published data (13), is that different cattle class I haplotypes carry varying numbers of expressed genes.

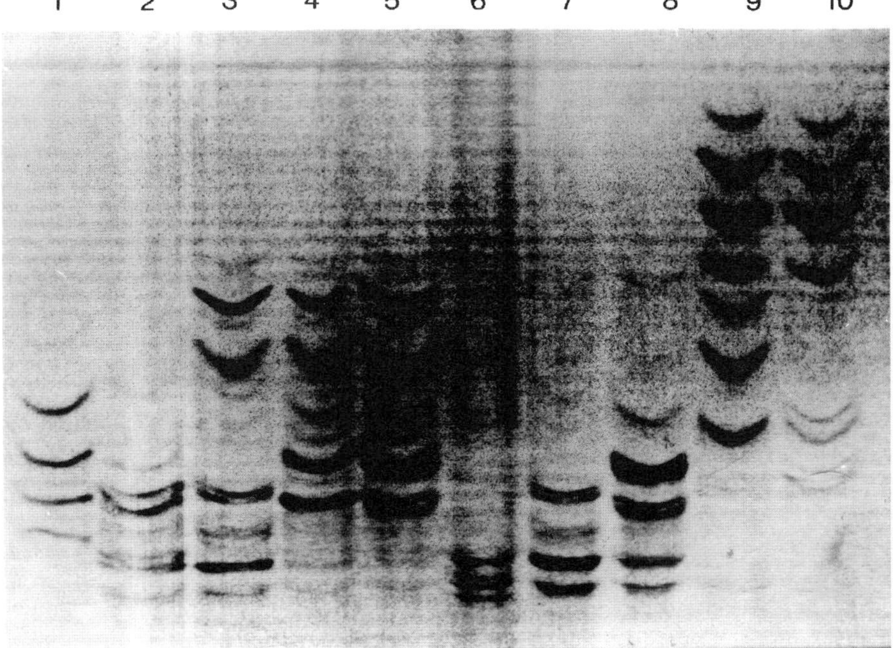

Figure 2. Autoradiograph of a 1D IEF gel. MHC class I heavy chains have been immunoprecipitated from ^{35}S-methionine-labelled lymphocytes from ten individuals.

Figure 2 shows a typical IEF gel, which illustrates some of these points. Each track represents immunoprecipitated class I heavy chains from a different individual. It is clear that there are many more bands in some tracks than others, and that the amount of radioactivity incorporated in individual bands is very variable. It is not clear whether this reflects true variation in the number of expressed products (and levels of expression), or variation in the affinity of the antibody for different molecules (the faint bands could, for example, represent immature products). Despite these problems it is possible to obtain useful information from this autoradiograph, even without supporting serology. For example, we can say that animals 4 and 5 have identical patterns, and therefore probably carry the same MHC class I haplotypes. Animal 3 shares two of these bands, so may perhaps share one or two alleles. Animals 9 and 10 share four bands, and so may share one haplotype. Obviously if these individuals were used for detailed molecular analysis, it would at a later stage be possible to match sequence data with banding patterns given the above information. These could then be confirmed by transfecting individual class I genes, and determining the pI of the product.

3.3 Molecular analysis

There are many methods available for molecular analysis of genes, and again it is necessary to ask what the information is needed for, before deciding how to proceed. Analysis of genomic DNA (e.g. RFLP analysis, mapping) will provide information on the MHC genes present, but will give little indication of which genes are functional or expressed. In the study described we were interested only in those genes coding for expressed class I molecules, which have an antigen-presenting role. MHC class I molecules are expressed at a reasonable level on resting lymphocytes, so these can provide a good source of mRNA. However, if obtaining blood is a problem, then transformed cell lines are a good alternative. In the study described, we used mRNA extracted from fresh peripheral blood lymphocytes, and also from *Theileria anulata*-transformed lymphocytes.

A standard procedure for identifying expressed genes would be to screen a cDNA library made from such mRNA. The problem with MHC class I is that in a heterozygous animal there will probably be at least four (and possibly more) different genes.

There will be multiple copies of each, and the most abundantly expressed could vastly outnumber the least expressed. Since we are assuming no prior knowledge about this, it would be necessary to go through a laborious procedure of checking each positive clone. In the example given, it was decided to obtain some information regarding the number and nature of expressed genes, and their relative frequency, by PCR analysis, prior to screening cDNA libraries.

3.3.1 PCR analysis

It is important to consider which areas of the MHC gene will be most informative, given that this is designed to be a quick screening procedure. The area chosen should be of a size that can be sequenced ideally on a single sequencing gel, i.e. in the region of 300 bp, and should be flanked by areas that are relatively non-polymorphic, such that primers can be designed which should amplify all of the class I genes present. *Figure 3* shows a representation of a class I gene. It was decided that amplifying the transmembrane (TM) and cytoplasmic (C) regions would be appropriate. In the case of human and mouse MHC class I, where many sequences are available, it can be seen that it would be relatively easy to assign an unknown sequence to a locus, based on this area where TM length and locus-specific amino acid residues are diagnostic.

Figure 4 shows the bovine class I sequences that were used to design the primers (18). mRNA is isolated from freshly prepared lymphocytes (or cultured transformed lymphocytes) using any appropriate commercially available kit, e.g. Dynal mRNA Direct kit (Dynal, Oslo). First strand cDNA synthesis is carried out using a cDNA synthesis kit, e.g. DNA synthesis system, BRL. (Note: general molecular biology procedures are described in detail in ref. 19.) PCR amplification can be performed directly on mRNA, but for reasons of stability it

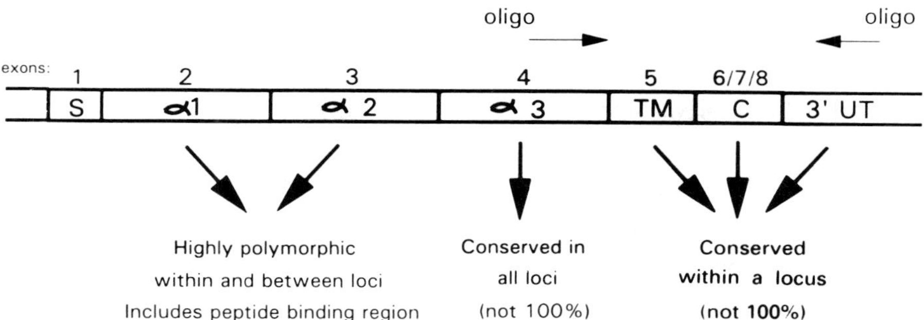

Figure 3. Predicted areas of sequence diversity and conservation within MHC class I genes (based on human and mouse).

```
                  ——————— OLIGO 2 ———————→
                                 Exon 4
Aw10      CCC CTC ACC CTG AAA TGG G ▒ AA CCT CCT CAG CCC TCC TTC CTC ACC ATG GGC ATC ATT GTT
BL3-6     ... ... ... ... .G. ... . ▒ .. ... ... ... ... ... ... ... ... ... ... ... ... ...
BL3-7     ... ..A ...  .. .G. ... . ▒ .. ... ... ... A.. ... ... ... .T. ... ... ... ... ...
KN104     ... ... ...  .. .G. ... . ▒ .. ... ... ... A.. ... ... ... ... ... ... ... ... ...

Aw10      GGC CTG GTT CTC CTC GTG GTC ACT GGA GCT GTG GTG GCT GGA GTT GTG ATC TGC
BL3-6     ... ... ... ... ... ... ... ... ... ... ... ... ... ..G ... ... ... ...
BL3-7     ... ... ... ... ... ..A ..* *** **. ... C.. ... ... ... .C. ... ... ..G
KN104     ..T ... ... ... .** *.T ... ... ... ... ... ... ... ... T.. ... ... ..G

                         Exon 5                                                Exon 6
Aw10      ATG AAG AAG CGC TCA ▒ GGT GAA AAA CGA GGG ACT TAT ATC CAG GCT TCA A ▒
BL3-6     ... ... ... ... ... ▒ ... ... ... G.. ..C .A. ... ... ... ... ... . ▒
BL3-7     .G. ... ... ... ... ▒ ... ... ... G.. C.. .TC ..C .C. ... ... G.. . ▒
KN104     ... ... ... ... ... ▒ ... ... ... G.. ... .A. ... ... ... ... ... . ▒

                                                                       Exon 7    Exon 8
Aw10      GC AGT GAC AGT GCC CAG GGC TCT GAT GTG TCT CTC ACG GTT CCT AAA G ▒ TG TGA
BL3-6     .. ... ... ... ... ... ... ... ... ... ... ... ... ... ... ... . ▒ .. ...
BL3-7     .. ... ... ... ... ... ... ... ... ... ... ... ... ... ... ... . ▒ .. ...
KN104     .. ... .C. ... ... .G. ... ... ... ... ... ... ... ... ... ... . ▒ .. ...

                     ←——— OLIGO 3 ———
          3' UT
Aw10      gacacctgccttcggggactgagtgatgcttcatcccgc
BL3-6     ....g.......at.......................a.
BL3-7     ....g.......gt.................g.....a.
KN104     ............at.........................
```

Figure 4. Four bovine MHC class I cDNA sequences (exon 4 → 3′UT) showing position of oligonucleotide primers 2 and 3.

is advisable to use cDNA, particularly if samples are to be stored, and re-amplified at a later date.

Protocol 1. PCR analysis

For each animal to be investigated use approximately 500 ng of cDNA in the PCR reaction. Always include a tube with no DNA as a control. Take extra care to ensure that no contamination occurs while setting-up the tubes because of the nature of the reaction any amount of contaminating DNA will render the results meaningless.

Equipment and reagents

- Thermal cycler
- Low melting point agarose
- 10 × reaction buffer[a]: 100 mM Tris, 40 mM $MgCl_2$, 0.5% Tween 20, 0.5% NP-40 pH 8.0
- 0.5 mM dNTP
- *Taq* DNA polymerase
- Phenol:chloroform
- Polynucleotide kinase

Method

1. To each tube on ice add DNA, 1 μM of each primer (final concentration), 10 μl 10 × reaction buffer, 0.5 mM each of dGTP, dATP, dTTP, dCTP. Finally add 2.5 U of *Taq* DNA polymerase, and make up the volume to 100 μl with DEPC treated water. Mix carefully, overlay with mineral oil, and place in a thermal cycler.
2. Programme the thermal cycler: 94 °C for 1 min, 50 °C for 1 min, 72 °C for 2 min, for thirty cycles; followed by a final cycle of 94 °C for 1 min, 50 °C for 1 min, 72 °C for 10 min. It may be necessary to adjust this according to the primers used in order to obtain optimal results.
3. After the PCR run analyse a fraction, e.g. 10 μl of the products on a 2% agarose gel, alongside appropriate size markers.
4. Following confirmation that the correct size of DNA has been amplified in each case, it is necessary to extract the remainder. Run the remaining 90 μl of PCR product on a 2% low melting point agarose gel. Excise the band using a scalpel blade, weigh it, and add water to 600 mg. Melt the agarose at 65 °C for 15 min, then extract once with phenol, once with phenol:chloroform, and once with chloroform. (Note: warm the phenol to avoid precipitating the agarose.) The amplified DNA can then be ethanol precipitated. There are commercial preparations available for extraction of DNA from agarose, e.g. GeneClean (Bio 101), which may give as good a yield of DNA.
5. The extracted DNA must then be treated with polynucleotide kinase in order to blunt-end ligate it into *Sma*1 cut M13 which has been treated with phosphatase to prevent self-ligation (19). Ethanol precipitate the kinased DNA.

[a] Note: most manufacturers of *Taq* polymerase will provide a reaction buffer (usually without magnesium), which can be substituted.

Figure 5 illustrates the procedure. If the primers have been successful, a mixture of all the different class I gene TM and C domains will have amplified. The different sequences should be present in approximately the same proportions as were present in the original mRNA. At this stage it is necessary to

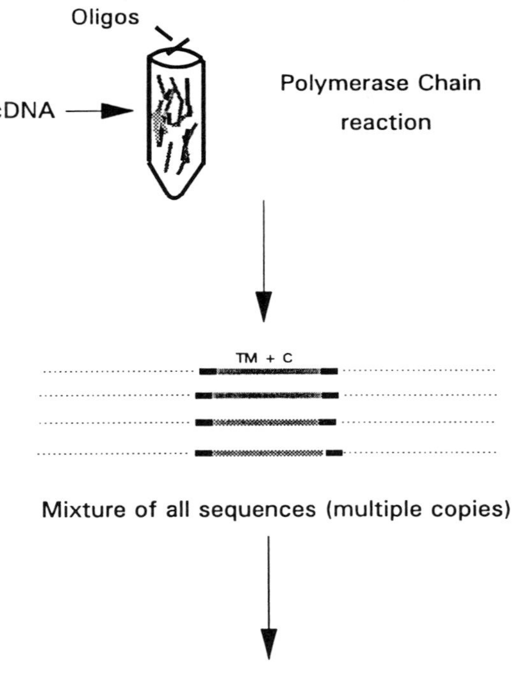

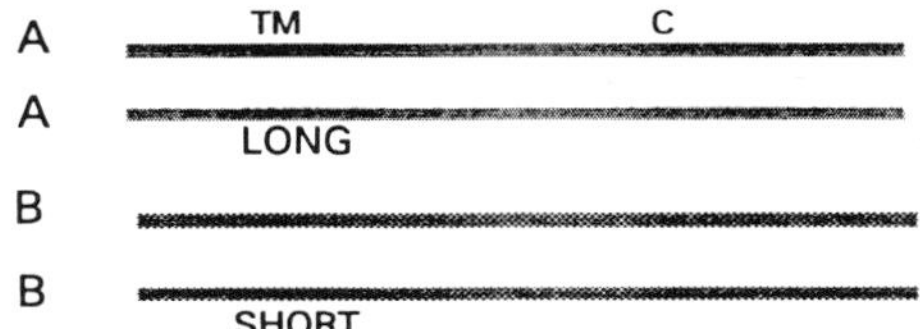

Figure 5. Strategy for analysis of expressed MHC class I genes = PCR amplification from cDNA, cloning of products into M13, and sequencing.

```
           Transmembrane region                       Cytoplasmic region

Aw10       EPPQPSFLTMGIIVGLVLLVVTGAVVAGVVICMKKRS      GEKRGTYIQASSSDSAQGSDVSLTVPKV
BL3-6      ...................................        ...G.N......................
D19.1      ....T................V....A..W..R..        ...G.N......................
D19.2      ...................................        ............................

BL3-7      ....T...I..........**.L...A..WR....        ...GRI.T..A.................
D19.3      ....T..............**.....A..WR....        ...GRI.T..A.................
D19.4      ....T...I..........**.....A..WR....        ....Q..T..A.G..D............
```

Figure 6. Four sequences obtained from one animal (D19) using the strategy illustrated in Fig. 5, compared to published sequences AW10, BL3–6 and BL3–7. Sequences are shown as amino acids to illustrate differences in TM length and possible locus-specific residues.

make a decision concerning how many sequences to examine. If, for example, one sequence was present at a level 100 times less than the others, then it is unlikely that it will be detected. Assuming that differences in levels of expression may be of the order of ten times, then sequencing 50 or 60 clones should be sufficient. In the case described approximately 60 clones were examined, and four different sequences were obtained at roughly equal levels (D19.1–D19.4) (*Figure 6*). A similar result was obtained with each animal investigated (18). In each case one or two additional sequences were found, which occurred at very low frequency. The implications of this will be discussed later in the section.

Protocol 2. Subcloning and sequencing

Equipment and reagents

- 2 × TY medium
- 2 × TY plates
- X-gal
- IPTG
- 100 mM $CaCl_2$
- M13
- Centrifuge

Method

1. Assuming a reasonable recovery of DNA (*Protocol 1*, step 5) (i.e. a visible pellet), resuspend it in 20 μl of water, and use 5 μl in the ligation reaction. Ligate overnight at 16°C into *Sma*1 cut and phosphatased M13 (mp18 or 19), 10 ng. (Standard ligation procedure as in ref. 19.) Always include controls:
   (a) No insert DNA, to check for background level of self-ligation.
   (b) Known blunt-ended insert, to ensure the ligation procedure is working.
   (c) Uncut M13 vector, to ensure viability of competent bacteria.
2. Make competent bacteria, e.g. TG1 or TG2 strain. These must be maintained on minimal agar plates. Pick a single colony and grow overnight at 37°C in 2 × TY medium. The following morning, add 0.5 ml of the culture to 50 ml of 2 × TY, and shake at 37°C until the OD_{260} is 0.2–0.6.

Place on ice for 15 min, spin at 250 *g* at 4°C for 15 min, place on ice for 10 min. Add 2 ml of 100 mM $CaCl_2$, resuspend extremely carefully. Keep on ice until ready for use. Competent bacteria can be frozen, but better results are usually obtained when they are freshly prepared.

3. Mix ligation with 100 μl competent bacteria. Leave on ice for 40 min, heat shock at 42°C for 2 min. Mix with 0.3 ml of overnight culture, 3 ml of melted top agar (at 42°C), 25 μl X-gal (20 mg/ml), 25 μl IPTG (20 mg/ml), and pour immediately on to dry 2 × TY plates. Grow overnight at 37°C.
4. White plaques signify the presence of an insert in the M13. Store plates at 4°C, and as convenient pick in the region of 60 plaques to prepare individual preparations of single-stranded template (19). The dideoxy method of Sanger *et al.* (20) can be used to obtain the sequence of these clones, with a sequencing kit such as USB Sequenase version 2, and an M13 primer. It is worth noting that there are several kits available now which use a non-isotopic approach for sequencing, e.g. Promega silver sequence kit.

These sequences should provide a significant amount of information. For example, they should show how many expressed genes are present (assuming expression at a reasonable level). Locus-specific characteristics may be apparent, e.g. differences in TM length, amino acid residues which always occur in certain positions at one locus. They may show that one locus product is expressed at significantly higher levels than the others. They will also provide a section of sequence which is unique to a particular allele (assuming that all alleles show differences in this region). This information can be very useful when screening cDNA libraries (see next section).

The two main drawbacks to this method are that because the primer sequences may be based on very limited information, they may prove to be inappropriate for amplifying some alleles, or even a whole locus. Because this is a short piece of DNA, major artefacts of the procedure should not be a problem. However, where a particular sequence occurs only once or twice, it is not possible to exclude this as an explanation, particularly if differences are minor (e.g. one or two nucleotides).

When a larger number of full-length or partial sequences become available, it is possible to modify the procedure to yield more information. The exact approach taken depends on the ultimate aim of the study. For example, primers can be designed which will amplify the whole peptide binding region, either from all class I sequences, or in a locus-specific manner. If sequences are needed for phylogenetic analysis, it could be more appropriate to amplify exons 4–8. In some cases this information will suffice, and it will not be necessary to go on to obtain full-length sequences for each allele.

3.2.2 Obtaining full-length sequences

There are two approaches to obtaining full-length MHC class I sequences. One is to screen a cDNA library using a probe derived from a known sequence (from the same or a different species), and the other is to use PCR amplification. There are a number of problems associated with the latter approach, mainly involving the need to design appropriate primers, and to avoid artefact (together with the associated necessity of sequencing multiple clones)—the advantages are speed and simplicity. The use of PCR to amplify full-length human MHC sequences is described in detail by Little and Parham (14). There have been a large number of recent publications describing cloning and sequencing of class I genes from a diverse range of species, and often a combination of these approaches is used to gain maximum information (e.g. 21, 22).

In the case of a species where there is little or no background information on the MHC, and where maximum data is required, it is advisable in the first instance to screen a cDNA library. When aiming to obtain full-length sequences from a cDNA library there are two important points to follow which improve chances of success. The first is to ensure that a high quality of mRNA is made, and that it is stored and handled in a way which minimizes chances of degradation. The second is to size fractionate the cDNA prior to making the library. There are various methods available for doing this—we have found two of them to be equally successful. The first is to pass the cDNA over a 1 ml Sephacryl S-400 column (Promega), which removes fragments of DNA less than approximately 500 bp. The second is to run all of the cDNA on a 1% TAE agarose gel alongside size markers. The DNA will appear as a smear, and it is then possible to select an appropriate size range, e.g. > 1 kb, and physically remove this section of gel. The piece of gel is re-inserted into a clear lane, and reverse electrophoresis is carried out, which results in the DNA compressing into a single band. This can be cut out, and the DNA extracted as described previously.

The choice of vector for making a cDNA library is generally dictated by personal preference, available expertise and reagents, and the chosen screening procedure (e.g. DNA probes, monoclonal antibodies). In the case described lambda gt10 was chosen.

Screening a cDNA library for MHC class I is straightforward. Any mammalian class I cDNA can be used as a probe, e.g. human, mouse, bovine, since the degree of similarity between different MHC class I genes is high. Following hybridization filters can be washed in 1 × SSC, 0.1% SDS, although stringency can be altered if it is thought necessary. The percentage of positive clones can be high (e.g. 0.2%), so assuming a reasonably sized library there could be hundreds of positive clones to investigate. Some of these will be too small (i.e. not full-length), and many will be identical, and it is therefore necessary to introduce further screening steps.

Initially a manageable number of clones should be chosen with which to proceed, and the first step is to screen twice more to ensure that they are, in fact, clones. It is then important to establish the size of the insert. This can be done by making a miniprep of lambda DNA and cutting out the insert with a suitable restriction enzyme. The digested product can then be run on an agarose gel alongside size markers. It may not be possible to predict the exact size of a full-length clone, but it is likely to be in the region of 1300 bp. At this stage it is advisable to concentrate on such clones, but the others should be stored carefully so that if necessary they can be investigated at a later stage. Shorter inserts, for example, do not always represent incomplete cDNAs—they may be alternatively spliced versions of a gene.

There are various ways of establishing which of the clones are identical. Since the TM and C domains of the major class I products have already been established by PCR, this information can be used for screening. The primers used for the initial PCR can be used again to amplify the same regions from each full-length clone (using the miniprep DNA). The amplified products are cloned into M13 as previously described, and sequence is obtained. In this way it is possible to very quickly find clones which represent the main expressed sequences. It is also a quick way of looking for additional sequences. If such primers are not available, the full-length cDNAs should be cloned directly into M13. Using an M13 primer it is then possible to obtain sequence from either end of the clone. These sequences should identify identical clones, and also can confirm that the cDNA is full-length.

The complete sequence of a clone is obtained with the use of sequencing primers—five pairs should be sufficient to sequence in both orientations. If there are sequences already available it may be possible to design these prior to beginning. If not, it is necessary to proceed in a stepwise fashion, i.e. sequence as far as possible with the M13 primer, and then use this sequence to design the next oligo, and so on along the clone (see protocols given in ref. 17).

Once some full-length sequences are available (to aid in primer design) it is often more convenient to examine additional haplotypes using PCR amplification. The problems inherent in amplification of full-length class I cDNAs are described in detail in Parham *et al.* (23). These include point mutations, and recombination between alleles, both of which can be partly overcome by optimizing conditions of amplification, and by sequencing multiple clones to obtain a consensus.

The exact nature and location of primers depends on information available. If amplified products are to be used for transfection, it is preferable to locate both primers outside the coding region. However, if no sequence from the untranslated regions is available, it is still possible to achieve full-length amplification (24). The protocol for full-length amplification, and subsequent cloning and sequencing, is essentially as described in *Protocols 1* and *2*, with the following modifications:

(a) PCR cycling conditions: 94 °C for 1 min, 56 °C for 3 sec, 72 °C for 2 min for 25 cycles, followed by one cycle of 94 °C 1 min, 56 °C 3 sec, 72 °C for 10 min.

(b) *Taq* extender™ PCR additive (Stratagene) is used in the PCR mix to reduce the incidence of point mutations and to increase the frequency of complete extension of products. The details of the conditions are to an extent dependent on the oligos used (length and composition), and may need to be altered accordingly.

(c) Blunt-end ligation as described in *Protocol 2* can be used for cloning, alternatively restriction enzyme sites can be incorporated into the primers to allow directional cloning.

(d) Single-stranded template production for full-length sequencing is greatly enhanced in terms of yield and purity if an extra purification step is introduced, e.g. Wizard™ M13 DNA Purification System (Promega).

(e) It is essential that multiple clones are sequenced, to obtain a consensus. If a clone is required for functional studies, e.g. for transfection, it is necessary to choose either a clone which carries the consensus sequence, or one where nucleotide differences do not result in a coding change.

Once a representative number of sequences has been obtained, or if working with an inbred population with limited MHC diversity, it is possible to begin to develop DNA-based methods for MHC typing (25, 26).

4. Further steps

Obtaining full-length MHC sequences does not necessarily give much information on expression or function. For example, if alternatively spliced or prematurely truncated cDNAs are found, does this mean that they are expressed, or have significance *in vivo*? One way to investigate this is to transfect the genes and examine the proteins that are produced.

Considerations at this stage are choice of vector and cell line, and whether to also transfect the correct β_2-microglobulin. Subsequent use of the transfectants generally dictates these decisions. Transfectants which express a single class I product can be used in immunization protocols to make allele-specific monoclonal antibodies. They can be used for *in vitro* studies of T cell responses. However, one of the most exciting uses to emerge in recent years is in the study of antigen processing and presentation by class I molecules.

If a large number of transfected cells is grown up in culture, it is possible to isolate the expressed class I molecules together with the pool of associated peptides. (This does depend on having a good monoclonal antibody which will immunoprecipitate the molecule in question, together with the wrong β_2-m if necessary.) It is then possible to analyse the peptide pool, and to obtain allele-specific peptide motifs, or even complete peptide sequences, from infected or uninfected cells (27, 28). Information of this kind is proving very useful in establishing the exact processes involved in peptide loading and transport, and in studies of T cell responses to specific diseases, e.g. HIV, malaria (29).

The ability to characterize the MHC genes carried in different populations provides information on genetic diversity, and can lead to identification of contributory factors in disease resistance. The information generated by such studies is essential for the development of preventative strategies, e.g. vaccines, genetic manipulation, which are now favoured over more traditional control methods for pathogens of domestic animals. In addition, such studies reveal vital information concerning the complex evolutionary processes involved in the generation and maintenance of a functional and effective MHC.

Acknowledgements

The work described was supported by the Biological Sciences and Biotechnology Research Council, UK. A major contribution to the practical work, and design of the figures, was made by Karen Staines. PCR protocols were optimized by Jo Pichowski.

References

1. Monaco, J. J. (1992). *Immunol. Today*, **13(5)**, 173.
2. Hughes, A. L. and Nei, M. (1993). *Immunogenetics*, **37**, 337.
3. Howard, J. (1992). *Nature*, **357**, 284.
4. Watkins, D. I., Chen, Z. W., Garber, T. L., Hughes, A. L., and Letvin, N. L. (1991). *Immunogenetics*, **34**, 185.
5. Kaufman, J. (1996). In *HLA and MHC: genes, molecules and function* (eds. M. Browning and A. McMichael), pp. 1–21. BIOS Scientific Publishers, Oxford.
6. Goddeeris, B. M., Morrison, W. I., Toye, P. G., and Bishop, R. (1990). *Immunology*, **69**, 38.
7. Ellis, S. A. (1994). *Eur. J. Immunogenet.*, **21**, 207.
8. Bernoco, D. (1991). *Anim. Genet.*, **22**, 477.
9. Bensaid, A., Kaushal, A., MacHugh, N. D., Shapiro, S. Z., and Teale, A. J. (1989). *Anim. Genet.*, **20**, 241.
10. Barnstable, C. J., Bodmer, W. F., Brown, G., Galfre, G., Milstein, C., and Williams, A. F. (1978). *Cell*, **14**, 9.
11. Bensaid, A., Kaushal, A., Baldwin, C. L., Clevers, H., Young, J. R., Kemp, S. J., *et al.* (1991). *Immunogenetics*, **33**, 247.
12. Terasaki, P. I., Bernoco, D., Park, M. S., Ozturk, G., and Iwahi, Y. (1978). *Am. J. Clin. Pathol.*, **69**, 103.
13. Ellis, S. A., Staines, K. A., and Morrison, W. I. (1996). *Immunogenetics*, **43**, 156.
14. Little, A. M. and Parham, P. (1993). In *Histocompatibility testing: a practical approach* (ed. P. Dyer and D. Middleton), pp. 159–90. OUP, Oxford.
15. Joosten, I., Oliver, R. A., Spooner, R. L., Williams, J. L., Hepkema, B. J., Sanders, M. F. *et al.* (1988). *Anim. Genet.*, **19**, 103.
16. Yang, S. Y. (1989). In *Immunobiology of HLA* (ed. B. Dupont), Vol. 1, pp. 309–31. Springer–Verlag, New York.
17. Hajek-Rosenmayr, A. and Doxiades, I. (1989). In *Immunobiology of HLA* (ed. B. Dupont), Vol. 1, pp. 338–57. Springer–Verlag, New York.

18. Ellis, S. A., Braem, K. A., and Morrison, W. I. (1992). *Immunogenetics*, **37**, 49.
19. Sambrook, J., Fritsch, E. F., and Maniatis, T. (ed.) (1989). In *Molecular cloning, a laboratory manual*, 2nd ed. Cold Spring Harbor Laboratory Press, New York.
20. Sanger, F., Nickel, S., and Coulsen, A. R. (1977). *Proc. Natl. Acad. Sci. USA*, **74**, 5463.
21. Grimholt, U., Hordik, I., Fosse, V., Olsaker, I., Endreson, C., and Lie, O. (1993). *Immunogenetics*, **37**, 469.
22. Betz, U., Mayer, W., and Klein, J. (1994). *Proc. Natl. Acad. Sci. USA*, **91**, 11065.
23. Parham, P., Adams, E., and Arnett, K. (1995). *Immunol. Rev.*, **143**, 141.
24. Pichowski, J. S., Ellis, S. A., and Morrison, W. I. (1996). *Immunogenetics*, **43**, 253.
25. Krausa, P., Barouch, D., Bodmer, J., and Browning, M. (1995). *Eur. J. Immunogen.*, **22**, 283.
26. Van Eijk, M., Stewart-Haynes, J., and Lewin, H. (1992). *Anim. Genet.*, **23**, 483.
27. Falk, K., Rotzsche, O., Stevanovic, S., Jung, G., and Rammansee, H.-G. (1991). *Nature*, **351**, 290.
28. Gaddum, R. M., Willis, A. C., and Ellis, S. A. (1996). *Immunogenetics*, **43**, 238.
29. Hill, A. V. S., Elvin, J., Willis, A. C., Aidoo, M., Allsopp, C. E., Gotch, F. M., *et al.* (1992). *Nature*, **360**, 434.

10

Techniques to identify the rules governing class II MHC–peptide interaction

J. HAMMER and F. SINIGAGLIA

1. Introduction

Major histocompatibility complex class II molecules (MHC) are highly polymorphic membrane glycoproteins that bind peptide fragments of proteins and display them for recognition by $CD4^+$ T cells. The binding capacity of any given peptide to an MHC class II molecule depends on the primary sequence of the peptide and on allelic variations of residues in the binding site of the MHC receptor. Advances in our knowledge of the structure of MHC molecules, and a large number of peptide and MHC class II structure–function studies have provided an increasingly detailed picture of how allelic variations influence peptide binding and presentation to $CD4^+$ T cells (1).

Here we review recent approaches we have used to define the structural requirements for the interaction between peptides and MHC class II molecules.

The use of M13 peptide display libraries for the selection of large pools of HLA-DR binding phage and the identification of position-specific anchor residues is described in Section 2. Section 3 outlines the use of position 1 (p1)-anchored designer peptide libraries for the refinement of phage-derived motifs. In particular, these libraries allow the identification of position-specific inhibitory residues. Positive and negative effects of amino acid residues on binding can be estimated, allowing a quantitative prediction of HLA-DR–peptide interactions. Finally, in the last section we present a brief outline of how data derived from MHC–peptide binding studies can be incorporated into a computer software for the identification of MHC class II binding regions in protein antigens.

2. Screening of phage display libraries

2.1 Filamentous phage and the principle of screening phage display libraries

The filamentous bacteriophage M13 is a single-stranded, male-specific DNA phage. It consists of a stretched-out loop of single-stranded DNA, sheathed in

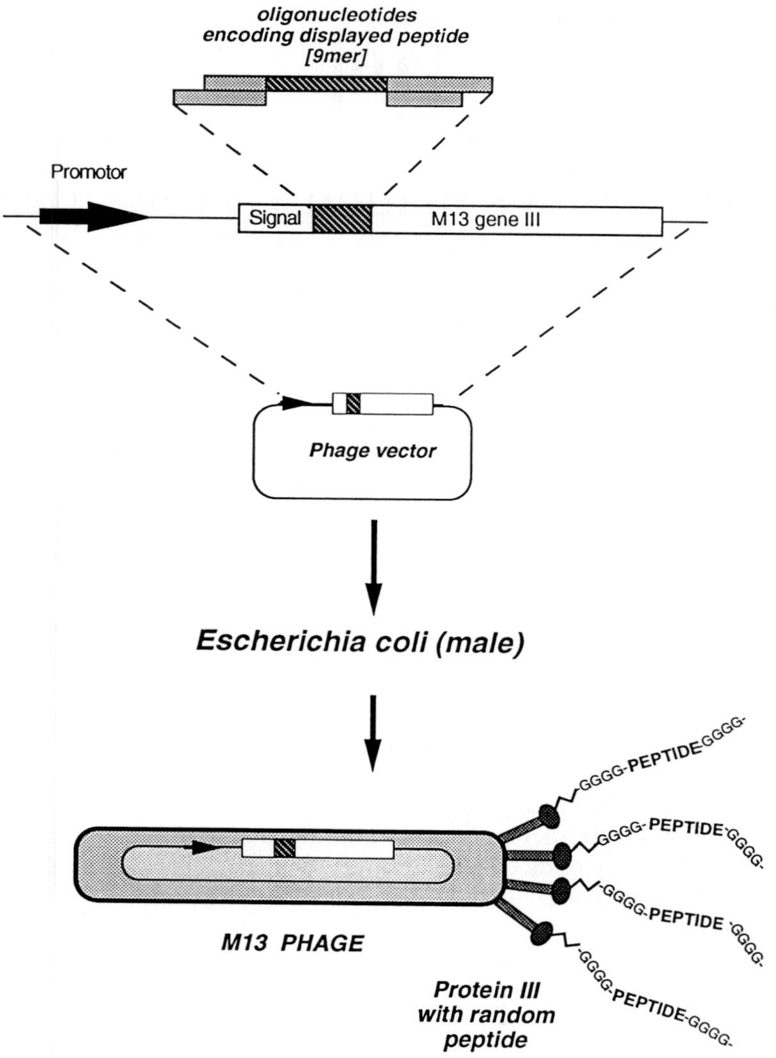

Figure 1. Construction of a bacteriophage displaying peptide epitopes. Oligonucleotides encoding peptides are inserted into the phage gene III and expressed as fusion molecules at the N terminus of minor phage protein III.

a tube composed of thousands of monomers of the major coat protein (gene VIII product). Minor coat proteins are found at the tips of the virion. One of the minor coat proteins, the protein III (product of gene III), attaches to the receptor at the tip of the F pilus of the host *Escherichia coli*.

The minor coat protein pIII folds into two domains. The carboxy terminal domain interacts with viral coat proteins and is required for viral assembly, while the amino terminal two-thirds of the protein forms a knob-like domain which projects away from the virion and is responsible for attaching to the F pilus.

Recently, the ability of filamentous bacteriophage to display foreign peptides on their outer surfaces has been demonstrated (2). Peptides fused to the NH_2 termini of the five copies of the pIII protein are displayed at one tip of each bacteriophage particle and have little or no effect on viral infectivity (*Figure 1*).

Phage display libraries are large collections of bacteriophage displaying different peptides on their surface (3–5). *Figure 2* summarizes the principle of

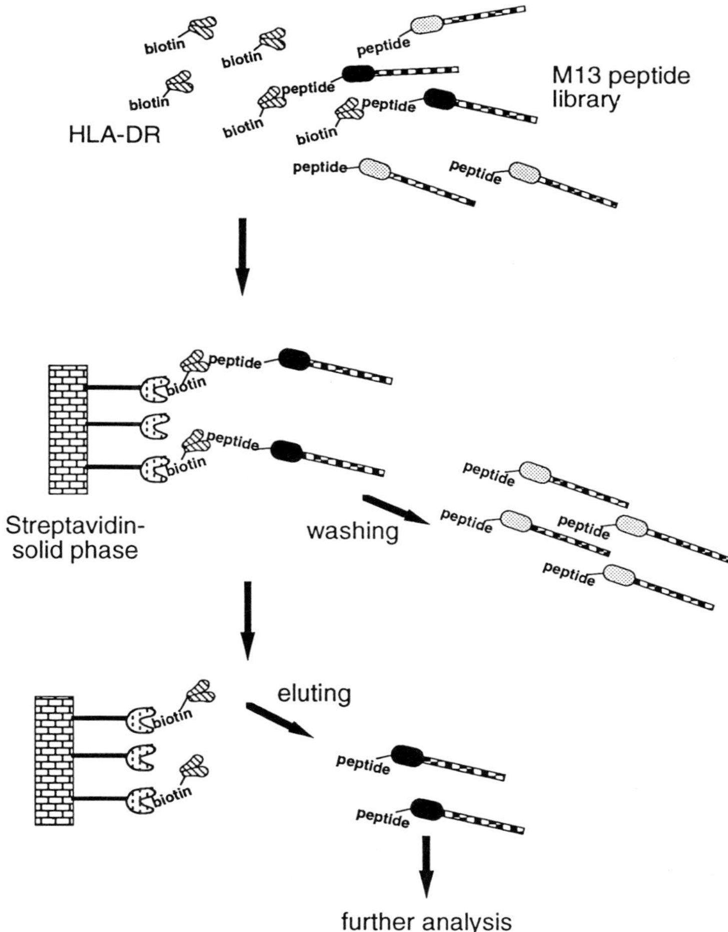

Figure 2. Schematic representation of the M13 peptide library screening procedure.

screening an M13 bacteriophage display library with MHC class II molecules. Purified biotinylated HLA-DR molecules are incubated with the library. Bacteriophage displaying peptides able to bind to DR molecules are in turn attached to a streptavidin solid phase via the strong biotin–streptavidin reaction, while unbound bacteriophage are washed away. Bound bacteriophage are then eluted with acid. Several rounds of screening, with intermediate amplifications, enrich for bacteriophage displaying peptides binding to the class II molecule. Sequencing of the corresponding phage DNA inserts reveals the amino acid sequences of the bound peptides.

2.2 Purification and biotinylation of HLA-DR molecules

HLA-DR molecules are affinity purified from human HLA-DR homozygous Epstein–Barr virus-transformed B cells. A three-step purification procedure is applied (*Protocol 1*). First, the cells are lysed and the nuclei are removed. Next, the membrane proteins are detergent solubilized. Finally, the DR molecules are isolated by affinity purification using a mouse monoclonal anti-human HLA-DR antibody (e.g. MAb L243). The yield of purified DR molecules from of 5×10^9 cells is between 2–4 mg.

The mouse monoclonal anti-human HLA-DR antibody is purified from mouse ascites by ammonium sulfate fractionation and DEAE anion exchange column chromatography. The purified antibody is cross-linked to protein A–Sepharose (Pharmacia) with dimethyl pimelimidate according to the manufacturers' instructions to give approximately 3 mg of antibody per millilitre of settled gel.

Protocol 1. Immunoaffinity purification of DR molecules

Reagents

- Lysis buffer: 1% (v/v) NP-40, 25 mM iodoacetamide, 1 mM phenylmethylsulfonyl fluoride (PMSF), 5 mM ε-amino-*n*-caproic acid, 10 μg/ml of each of soybean trypsin inhibitor, antipain, pepstatin, leupeptin, and chymostatin, in 0.05 M sodium phosphate buffer, pH 7.5, containing 0.15 M NaCl
- Neutralizing solution: 1 M Tris–HCl pH 6.8
- Wash buffer 1: 50 mM Tris–HCl pH 8, 0.15 M NaCl, 0.5% NP-40, 0.5% sodium deoxycholate (DOC), 10% glycerol, 0.03% NaN_3
- Wash buffer 2: 50 mM Tris–HCl pH 9, 0.5 M NaCl, 0.5% NP-40, 0.5% DOC, 10% glycerol, 0.03% NaN_3
- Wash buffer 3: 2 mM Tris–HCl pH 8, 1% octyl-β-D-glucopyranoside, 10% glycerol, 0.03% NaN_3
- Elution buffer: 50 mM diethylamine–HCl pH 11.5, 1% octyl-β-D-glycopyranoside, 0.15 M NaCl, 1 mM EDTA, 10% glycerol, 0.03% NaN_3
- Phosphate-buffered saline (PBS)

Method

1. Harvest 5×10^9 human HLA-DR homozygous Epstein–Barr virus-transformed B cells by centrifugation and wash twice with PBS.
2. Resuspend cell pellet to a cell density of 10^8/ml in lysis buffer and incubate on ice for at least 60 min.

3. Clear the lysate of nuclei and debris by centrifugation at 27 000 *g* for 30 min. If not used immediately, cells lysates can be stored at −70 °C.
4. Add 0.2 vol. of 5% DOC to the post-nuclear supernatant and mix for 10 min.
5. Centrifuge at 100 000 *g* for 2 h and filter the supernatant through a 0.45 μm pore-size membrane.
6. Mix the lysate with 5 ml (settled volume) of protein A–Sepharose coupled anti-human HLA-DR antibody and rotate at 4 °C for at least 3 h.
7. Wash the gel mixture twice with ten volumes of wash buffer 1 and transfer it into a column.
8. Wash the column using a flow rate of 0.5–1 ml/min at 4 °C with a least 20 column volumes of wash buffer 1, five column volumes of wash buffer 2, and five column volumes of wash buffer 3.
9. Elute the DR antigens with two column volumes of elution buffer, collect 1 ml fractions, and neutralize fractions with 100 μl of neutralizing solution.
10. Analyse aliquots of each fraction by SDS–PAGE. Pool fractions containing most of the DR antigens, aliquot, and keep them frozen at −70 °C until used.

After the binding of bacteriophage to HLA-DR molecules it is necessary to remove the DR/bacteriophage complexes from the phage mixture. The biotinylation of DR molecules allows the attachment of the DR-bound phage to a streptavidin solid phase, while unbound bacteriophage are washed away. The procedure described in *Protocol 2* allows biotinylation of approximately 80% of HLA-DR molecules. Thus, 80% of the DR molecules binding a peptide can be precipitated by streptavidin–agarose.

Protocol 2. Biotinylation of DR molecules

Equipment and reagents

- Biotinylation buffer: 0.25 M $NaHCO_3$ pH non-adjusted, 0.2% Nonidet P-40
- BDR-buffer: 50 mM Tris–HCl pH 7.5, 150 mM NaCl, 2 mM EDTA, 0.2% Nonidet P-40
- PD-10 columns, Sephadex G-25M (Pharmacia)
- Biotin-XX-NHS (Calbiochem): 25 mM in dimethylformamide
- Microsep Microconcentrators, 30 kDa cut-off (Scan AG)

Method

1. Exchange the buffer of 1 mg affinity purified HLA-DR antigen solution (*Protocol 1*) into biotinylation buffer by gel filtration using two PD-10 columns sequentially, according to the manufacturers' instructions.

Protocol 2. *Continued*

2. Reduce the volume of the HLA-DR solution to 1 ml by ultradialysation with Microsep Microconcentrators.
3. Add to the concentrate 8 μl of a 25 mM solution of biotin-XX-NHS and rotate for 1 h at room temperature.
4. Remove the excess biotin-XX-NHS by gel filtration using a PD-10 column equilibrated with BDR-buffer.
5. Concentrate the biotinylated DR antigens by ultradialysation (see above), aliquot, and keep them frozen at −70 °C until used.

2.3 Construction of an M13 peptide display library

2.3.1 Construction of an M13 library vector

Phage peptide libraries displayed by protein III have been constructed in fd and M13 bacteriophage (3–6). We suggest using the commercially available M13mp19 vector for constructing the M13 library vector. It gives the advantage of blue/white selection which simplifies the MHC class II–phage binding assay (*Protocol 4*). *Figure 3* shows the construction of the M13 library vector. A β-lactamase gene is placed into the *Eco*RI and *Bam*HI sites of the M13mp19 polylinker to confer ampicillin resistance to the host, which allows the propagation of bacteriophage, like plasmids, independently of phage function. In addition, the disruption of the β-galactosidase gene fragment by the β-lactamase gene creates white plaques in X-gal indicator plates, while M13mp19, used as reference phage in the MHC class II–phage binding assay, produces blue plaques. An *Ecl*XI site is created by site-directed mutagenesis in a region of the gene III, which encodes the post-processing NH_2 terminus of protein III (*Figure 3*). A stuffer fragment is designed containing two common restriction enzyme sites to allow directional cloning, e.g. *Sac*I and *Kpn*I, which were removed from the M13mp19 polylinker by the insertion of the β-lactamase gene. The stuffer fragment is cloned into the *Ecl*XI site of the M13 construct. The part of the stuffer fragment which is flanked by the *Sac*I and *Kpn*I sites is then either replaced by random oligonucleotides encoding the peptide library (see Section 2.3.2 and *Protocol 3*), or by oligonucleotides encoding known peptide sequences (see Section 2.4).

2.3.2 Construction of a random phage peptide library

We chose random peptide inserts which were nine amino acid residues in length and flanked by four glycine spacers on each side. *Figure 4* illustrates the construction of the random peptide region. A random oligonucleotide flanked by the spacer encoding region is hybridized to two 'half-site' oligonucleotides to form cohesive termini complementary to the *Sac*I and *Kpn*I sites in the M13 library vector. The hybridized structure is then ligated to the *Sac*I/*Kpn*I cut replicative form DNA (RF) of the M13 library vector to produce a double-

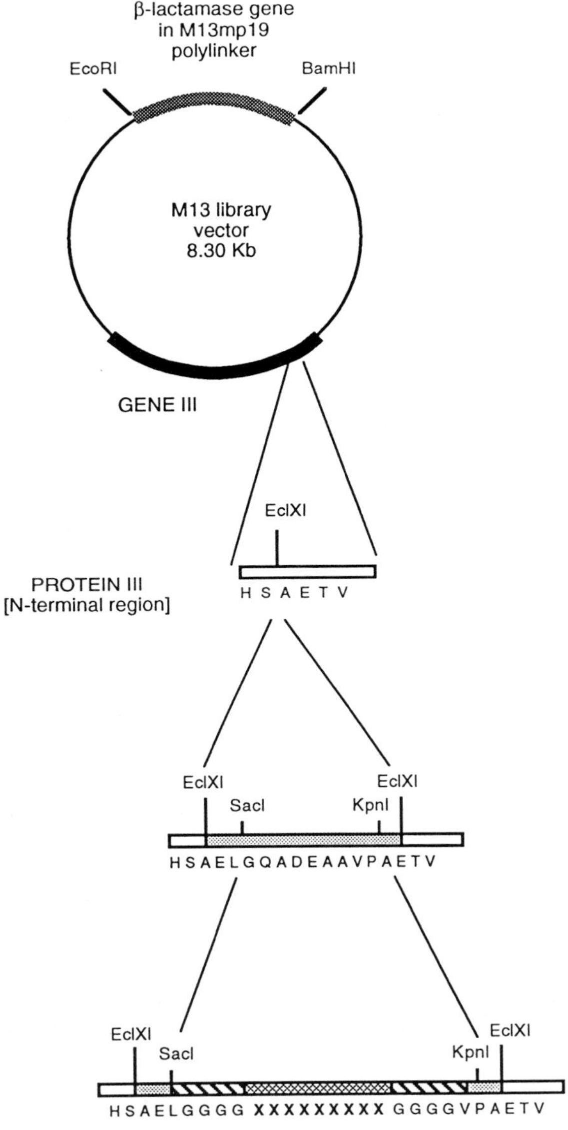

Figure 3. Construction of an M13 peptide display library vector.

stranded circular molecule with a small, single-stranded gap. Next, the recombinant molecules are transformed into SCS1 *E. coli* cells to produce an M13 peptide display library. *Protocol 3* presents an outline of the library construction.

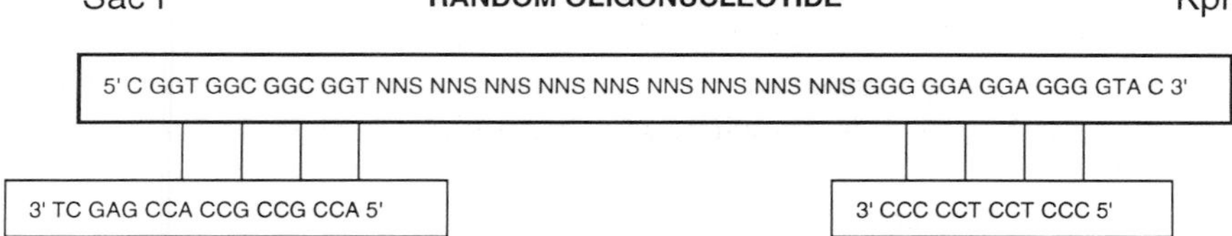

Figure 4. Generation of a random peptide for display on M13 bacteriophage.

The random oligonucleotide region does not contain a termination codon when transformed into SCS1 *E. coli* cells. The codon NNS (where N is a mixture of G, A, T, and C, and S is a mixture of G and C) encodes all 20 amino acids but will not produce the two termination codons UAA and UGA. The third termination codon UAG codes for the glutamine in SCS1 and XL1Blue cells because of an amber suppressor mutation ('supE') in these strains. In contrast to the fd-tet phage, which was first used for the construction of peptide displaying phage (2), a frame shift mutation cannot be inserted into the stuffer fragment of the M13 library vector, because it blocks the assembly of the virion and leads to the accumulation of phage DNA and gene products, which eventually kills the host. However, using a stuffer fragment without a frame shift mutation in M13mp19-derived phage may result in a higher background of phage without peptide inserts. Therefore great care has to be taken to digest the library vector to completion, and to entirely remove the stuffer fragment and single-stranded DNA by gel electrophoresis.

Protocol 3. Construction of M13 peptide display libraries

Reagents

- Electrocompetent *E. coli* SCS1 cells (SCS1 cells are obtained from Stratagene)
- Random oligonucleotide, phosphorylated (see Section 2.3.2)
- Two half-site oligonucleotides, phosphorylated (it is important that the oligonucleotides do not cross-hybridize with each other, especially if similar spacers are used on each site)
- M13 library vector (see Section 2.3.1)
- 10 × annealing buffer: 200 mM Tris–HCl pH 7.5, 20 mM $MgCl_2$, 500 mM NaCl
- SOC media: 50 ml LB (Luria–Bertani) media plus 125 μl 1 M KCl, 50 μl 1 M $MgCl_2$, 500 μl 1 $MgSO_4$, 1 ml 1 M glucose
- Components and protocols not listed here can be found in ref. 7.

Method

1. Digest to completion approx. 250 μg M13 library vector DNA with *Sac*I and *Kpn*I (Secton 2.3.1) and load digestion mixture on a 0.8% ultra pure agarose gel (gel volume: 400 ml, size: 35 cm × 20 cm). Run gel overnight with 30 V. This step removes undigested ssDNA and the stuffer fragment.
2. Cut out the digested RF-DNA band and elute by electroelution. Concentrate DNA with 1-butanol to a final volume of 400 μl. Extract twice

with phenol:chloroform (1:1), once with chloroform, and precipitate with ethanol. Wash and resolve DNA pellet in TE buffer.

3. Anneal the oligonucleotide encoding the random peptide library with the two half-site oligonucleotides in a ratio of 1:1:1 by mixing the following:
   - 50 mM half-site oligo 1 solution 40 μl
   - 50 mM half-site oligo 2 solution 40 μl
   - 50 mM random oligo solution 40 μl
   - 10 × annealing buffer 13 μl

   Transfer annealing mixture in an Eppendorf tube to a heating block at 95 °C and switch power off. Remove the tube as soon as the block is below 30 °C. Keep the annealed oligonucleotides on ice.
4. Ligate annealed oligonucleotides with digested RF-DNA in an estimated 5:1 ratio. Extract ligation mix with phenol:chloroform (1:1), and chloroform, and precipitate with ethanol. Resolve pellet in TE. The final concentration of the purified ligation mix should be at least 100 ng/μl.
5. Prepare electrocompetent *E. coli* SCS1 cells and make 220 μl aliquots. The transformation efficiency should be in the range of 10^9 with the pUC19 vector.
5. For transformation add around 1 μg of purified ligation mix to one aliquot of electrocompetent SCS1 cells (220 μl) placed on ice. For each transformation, transfer 40 μl of this mixture to an electroporation cuvette (Bio-Rad, 0.2 cm) and use the following parameters for the electroporation: 25 mF, 2.5 kV, 200 Ohm, cuvette holder and cuvettes ice-cold. After each electroporation immediately combine the transformed cells with 2 ml pre-warmed SOC media and incubate for 1 h at 37 °C and 220 r.p.m.
7. After incubation combine all five transformations in a Falcon tube. Plate 1 μl of this mixture on an agar plate containing 20 μg/ml ampicillin and grow overnight to determine the number of independent clones. The rest of the 10 ml mixture is transferred to 500 ml prewarmed LB/ampicillin media (20 μg/ml) and incubated for at least 24 h at 37 °C and 220 r.p.m. to amplify the library.
8. Harvest the amplified library by three polyethylene glycol precipitations of the phage supernatant and determine the titre of the amplified library.
9. With 1 μg ligation mix, 220 μl competent cells, and five transformations one should obtain around five million independent clones. To synthesize larger libraries, synthesize several small libraries as described above and combine them afterwards. The small libraries should be mixed in a way that no clones of the final library is under- or overrepresented. Therefore, the titre before and after amplification of every small library has to be measured to determine the right ratio.

2.4 Optimization of class II binding to bacteriophage

Since different MHC class II alleles require different peptide binding conditions, optimization experiments are required for each MHC class II allele prior to screening. To optimize binding between peptide displayed phage and class II molecules, peptides binding with high affinity (K_d within 10–100 nM) to the corresponding class II allele can be placed at the NH_2 terminus of protein III (*Figure 5*) and used in the MHC class II–phage binding assay (*Protocol 4*, *Figure 5*). Binding of M13 phage displaying peptides to DR molecules is determined by their enrichment as compared to the reference phage M13mp19. The enrichment factor is determined by mixing both types of phage and plating the initial mixture and the eluates obtained after binding to DR molecules on X-gal indicator plates. An enrichment of white (bacteriophage displaying peptides) versus blue plaques (M13mp19) indicates peptide-dependent binding.

An important parameter to be considered in binding studies is the pH of the binding buffer. *Figure 6a* demonstrates how the pH influences phage–DRB1*0401 binding. The crucial role of the pH has also been demonstrated in the specific selection of DR binding phage from large phage display libraries (8). The pH used for washing the streptavidin solid phase is equally important for the enrichment of DR-specific phage (*Figure 6c* and *d*). Identical pH conditions either reduce the number of non-specific reference phage in the case of DRB1*0101 molecules (*Figure 6c*), or prevent most of the binding of specific phage in the case of DRB1*0401 molecules (*Figure 6d*).

In comparison to other receptors used in screening phage display libraries, we generally find a rather low enrichment for phage binding to DR molecules (*Figure 5*). This finding could be due to a specific interaction of DR molecules with sites of the phage surfaces other than the peptide–insert region. This is best illustrated by the fact that the HA peptide, known to bind to the MHC

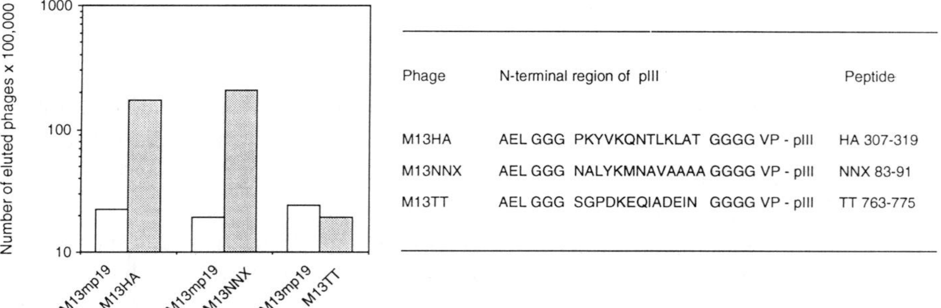

Phage	N-terminal region of pIII	Peptide
M13HA	AEL GGG PKYVKQNTLKLAT GGGG VP - pIII	HA 307-319
M13NNX	AEL GGG NALYKMNAVAAAA GGGG VP - pIII	NNX 83-91
M13TT	AEL GGG SGPDKEQIADEIN GGGG VP - pIII	TT 763-775

Figure 5. MHC class II–phage binding assay. Binding of M13 constructs is determined by their enrichment as compared to the reference phage M13mp19. For M13 constructs, the phage input is one billion, and for M13mp19, two billion.

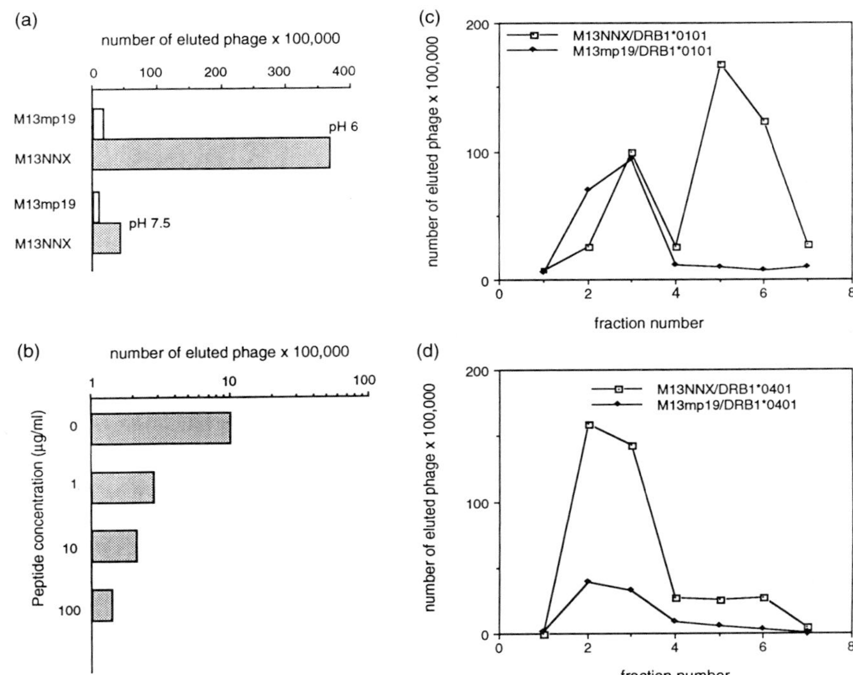

Figure 6. Optimization of MHC class II binding to bacteriophage. (a) MHC–phage interaction depends on the pH of the binding buffer. The phage M13NNX binds better to DRB1*0401 at pH 6 than at pH 7.5. The NNX 83–91 peptide sequence has been reported in ref. 6. (b) Background binding is DR-specific. The interaction of M13mp19 phage with DRB*0101 is specifically competed by the DRB1*0101 binding peptide HA 307–319 but not by the AChR 4–15 peptide (data not shown). (c) and (d) Different effects of an additional pH 3 wash step on the number of eluted phage. 10^9 phages displaying the NNX peptide known to bind strongly DRB1*0101 (upper panel) or DRB1*0401 (lower panel). After eight washes in binding buffer, the remaining phage were eluted as follows: fraction 1 and 3, binding buffer; fraction 2 and 4, pH 3 buffer; fraction 5–7, elution buffer. Phage numbers were determined by plating the mixture on X-gal indicator plates.

binding cleft, is able to inhibit the background binding of M13mp19 reference phage (*Figure 6b*).

2.5 Library screening

The screening of M13 display libraries with HLA-DR molecules is essentially performed as the MHC class II–phage binding assay described above. However, since only a small percentage of the library phage will display peptides binding to DR molecules, several rounds of screening with intermediate amplification steps are necessary to enrich for HLA-DR binding phage (*Protocol 5*). We perform up to five binding and amplification steps per screening. The binding capacity of the libraries can be monitored after each round of screening by the MHC class II–phage binding assay. Instead of using phage displaying known peptides, aliquots of amplified and DR-selected libraries are mixed with M13mp19 reference phage and the enrichment factor is determined as described in *Protocol 4*.

Protocol 4. MHC class II–phage binding assay

Reagents

- M13mp19 phage reference phage (Stratagene)
- Biotinylated HLA-DR molecules (see *Protocol 2*)
- Streptavidin on 4% beaded agarose (Sigma)
- BSA blocking solution: 5 mg/ml BSA fraction V, in phage binding buffer
- Phage binding buffer: 150 mM NaCl, 2 mM EDTA, 0.2% Nonidet P-40, 1 mM phenylmethylsulfonyl fluoride (PMSF), citrate–phosphate buffer—optimal pH is HLA-DR allele specific (see Section 2.4)
- Elution buffer: 0.1 N glycine–HCl pH 2.2, 1 mg/ml BSA fraction V

Method

1. Mix up to one billion M13 bacteriophage displaying peptides of known or random sequences with a similar number of M13mp19 bacteriophage as a reference, and incubate with 10–50 pmol biotinylated DR1 in phage binding buffer.
2. To reduce non-specific binding, wash streptavidin on 4% beaded agarose twice with 10 vol. BDR-buffer (1 vol. = volume of settled streptavidin–agarose), incubate for 1 h with 10 vol. BDR-buffer containing 5 mg/ml BSA, and wash subsequently three times with 10 vol. BDR-buffer.
3. After at least 24 h incubation at room temperature, add 250 μl BSA blocked streptavidin on 4% beaded agarose (25 μl settled volume) and incubate for 10 min.
4. Purify the M13 phage/DR complexes by washing the solid phase several times with BDR-buffer.
5. Elute the bacteriophage with 1 ml elution buffer for 10 min and neutralize with 30 μl 2 M Tris base.
6. Determine the ratio of the bacteriophage displaying peptides to M13mp19 bacteriophage in both the initial mixture and the eluates by plating aliquots on X-gal indicator plates (7). An enrichment of white (bacteriophage displaying peptides) versus blue plaques (M13mp19) indicates peptide-based binding.

We observed that the more rounds of screening are performed, the higher is the chance of isolating phage displaying identical peptides. Because the identification of DR peptide binding motifs requires the alignment of large numbers of peptides derived from independent DR binding phage we usually select, for sequencing, phage from the DR-selected library first exhibiting a clear enrichment over the M13mp19 reference phage.

Another problem frequently encountered by screening M13 peptide libraries with class II molecules is the enrichment for non-MHC-specific bacteriophage. This background binding could be due to phage binding to streptavidin, to biotin, or to agarose. Indeed, whenever binding conditions are non-optimized, we observe enrichment for non-specific phage displaying peptides carrying several Trp residues. When tested, these phage bind to any biotinylated protein or to the streptavidin solid phase.

Protocol 5. Screening of phage display libraries with DR molecules

Equipment and reagents

- Amplified M13 peptide display library (see *Protocol 3*); we used libraries consisting of 20 million independent clones.
- *E. coli* XL1Blue plating bacteria (optical density = 0.5) (XL1Blue can be obtained from Stratagene): plating bacteria are prepared as described (7)
- Biotinylated HLA-DR molecules (see *Protocol 2*)
- BSA blocked streptavidin on 4% beaded agarose (see *Protocol 4*)
- Phage binding buffer (*Protocol 4*), BDR-buffer (*Protocol 2*), elution buffer (*Protocol 4*)
- Phage storage buffer (PSB): 50 mM Tris–HCl pH 7.5, 150 mM NaCl
- PEG solution: 20% polyethylene glycol (PEG 8000) in 2.5 M NaCl
- PD-10 columns and Microsep Micro-concentrators (see *Protocol 2*)

Method

1. Incubate approximately 1×10^{10} bacteriophage of an amplified M13 peptide display library with 5 μg biotinylated HLA-DR molecules in phage binding buffer at room temperature for 48 h (total volume: 60 μl).
2. Add 500 μl BSA blocked streptavidin on 4% beaded agarose in BDR-buffer (total 50 μl settled volume) to the library/HLA-DR incubation mix and rotate for 10 min at room temperature.
3. Purify the M13 phage/HLA-DR complexes by washing the streptavidin solid phase eight times with BDR-buffer using 1 ml per wash step. Additional washing steps are optional (see Section 2.4).
4. Elute the bacteriophage with 1 ml elution buffer for 10 min and neutralize with 60 μl 2 M Tris base.
5. For amplification, infect 1 ml of *E. coli* XL1Blue plating bacteria with 300 μl eluted bacteriophage and transfer cells to 7 ml LB medium in a 25 cm^2 tissue culture flask. After 1 h incubation (37 °C at 200 r.p.m.), add ampicillin to a final concentration of 20 μg/ml and incubate for further 20 h at 37 °C and 200 r.p.m.
6. For harvesting and purification of the bacteriophage, transfer the cell/phage mixture to 15 ml tubes (Falcon, polypropylene) and centrifuge at 2500 *g* for 25 min at room temperature. Transfer 5 ml of

Protocol 5. *Continued*

supernatant into a new tube, add 1 ml PEG solution, and mix by inverting the tube several times. Recover the precipitated bacteriophage particles after 10 min incubation at room temperature by centrifugation using the same conditions as above. Aspirate the supernatant, centrifuge for 30 sec, and remove any residual supernatant. Resolve the phage pellet in 1.2 ml PSB and transfer it into a 1.5 ml tube (Eppendorf). Add 200 μl PEG solution and recover the phage particles by centrifugation. Resolve the pellet in 500 μl PSB and store at 4 °C until used.

7. Use an aliquot of this phage solution (approx. 1×10^{10} phage particles) for the next round of screening and amplification (step 1 to step 6).
8. Phage from the first DR-selected library which exhibits a definite enrichment over M13mp19 phage should be isolated to sequence the peptide coding region (7).

2.6 Analysis and interpretation of results

To investigate the structural characteristics of peptides capable of binding to DR molecules, the peptide encoding region of a large number of DR-selected phage is sequenced. For the identification of anchor positions, an unbiased alignment is performed using the *PILEUP* program of the GCG (Genetics Computer Group Inc.) program package. As an example, the alignment of DRB1*0401 selected peptide sequences and the identification of the major anchor positions are shown in *Figure 7*.

If the alignment of phage-derived sequences does not reveal any peptide binding motifs, it is necessary to determine the percentage of DR binding peptides within the selected peptide pool. This should be done by testing synthetic peptides, based on the phage sequences, for their capacity to bind the corresponding DR molecules. This test is performed with the spin column peptide binding assay (*Protocol 6*). *Figure 8a* illustrates results obtained with the spin column binding assay. Only if a high percentage of the synthetic peptides binds to the corresponding class II allele is it justified to sequence a larger DR-selected peptide pool.

If alignment of phage-derived sequences reveals anchor positions, anchor-addition and anchor-shifting experiments may subsequently be performed to confirm their role in peptide–MHC interactions (*Figure 8b* and *c*). Anchor-substitution binding experiments should also be carried out to demonstrate the effect of MHC allele-specific anchors (8).

How representative of the naturally processed peptide pool are the HLA-DR peptide binding motifs derived from phage display libraries? Sequences of naturally processed peptides associated with several different HLA-DR alleles have been identified (for review see ref. 1). Although the peptides described in these studies in general conform to the binding motif identified with

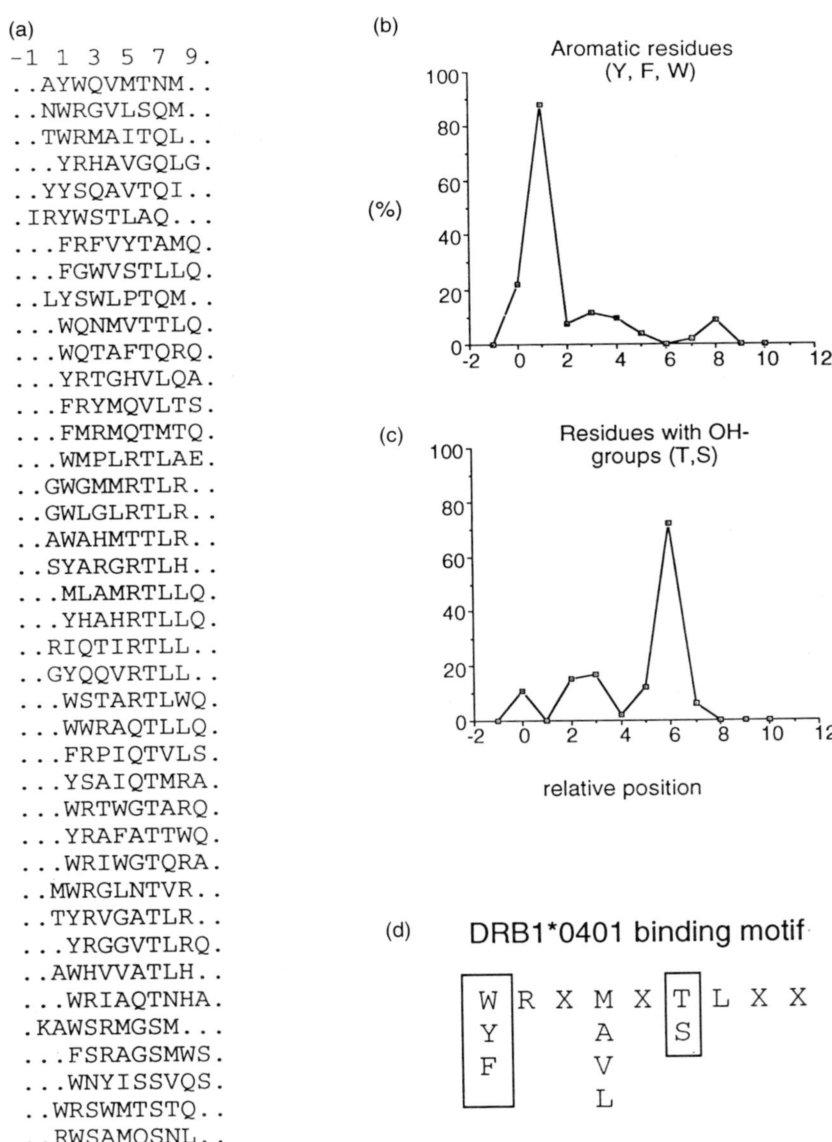

Figure 7. Identification of anchor positions within HLA-DR selected peptide pools. (a) Alignment of peptides displayed by DRB1*0401-selected phage. (b) and (c) Identification of the major anchor positions at p1 and p6. (d) The DRB1*0401 peptide binding motif as derived from screening M13 display libraries (8, 9).

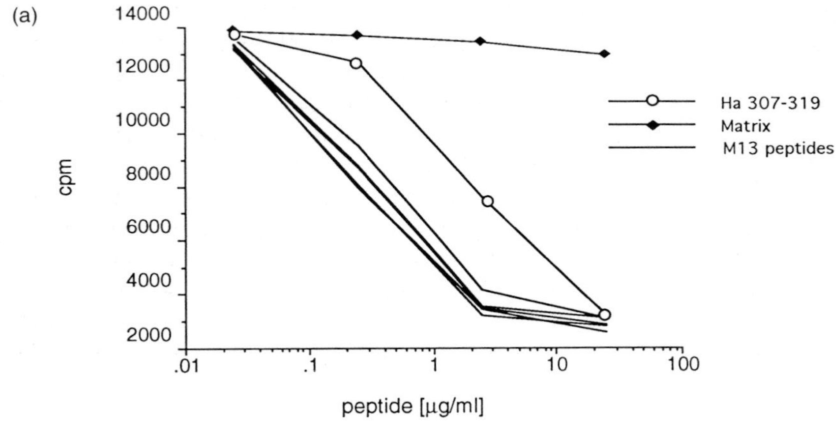

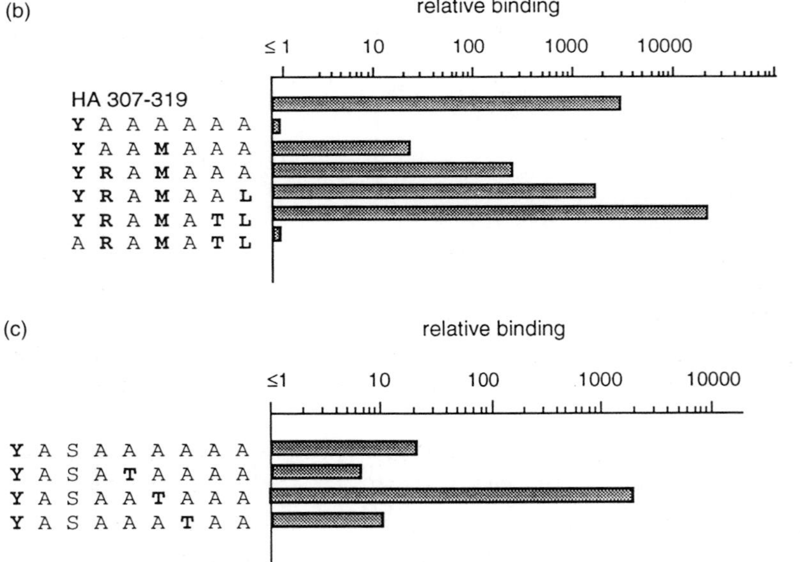

Figure 8. Verification of results obtained from screening M13 display libraries (a) Spin column peptide binding assay: five out of five peptides based on sequences corresponding to those of DRB1*0401-selected phage bind to the DRB1*0401 molecule. (b) Anchor-addition experiments: the effect of anchor residues on peptide affinity is determined by their addition into Ala-based designer peptides. (c) Anchor-shifting experiments: the position specificity of anchor residues is tested by shifting anchor residues in designer peptides.

the phage library screening, some minor differences could be found. Position 1 for example consisted mainly of aromatic residues in DR-selected peptide pools (*Figure 7*), whereas aliphatic residues can also be found in this position in naturally processed peptides. Indeed, in binding experiments with designer peptides, in addition to aromatic, also hydrophobic residues at p1 enable peptide binding to DR molecules, albeit at lower affinity (*Figure 9*).

The screening of phage display libraries with MHC class II molecules is certainly a powerful way to identify anchor position as potential contact sites of amino acid side chain interaction with the MHC class II binding cleft. Besides anchor residues, other sites have, however, been shown to play an important role by preventing effective interaction with the MHC molecule. The following section provides an approach for the identification of these inhibitory residues.

Protocol 6. Spin column peptide binding assay

Equipment and reagents

- 2 × binding buffer: 2% NP-40, 2 mM PMSF, 2 mM EDTA, 20 mg/ml of each of the following protease inhibitors: soybean trypsin inhibitor, antipain, pepstatin, leupeptin, and chymostatin, 0.15 m NaCl, 50–100 mM citrate–phosphate buffer (optimal pH needs to be determined)
- Immunoaffinity purified DR antigens (see *Protocol 1*)
- Bio-Spin columns (Bio-Rad)

Method

1. Mix 25 μl (1–5 μg) of immunoaffinity purified DR antigens, 25 μl of 2 × binding buffer, 5 μl (5 ng) of ^{125}I-labelled peptide, and 5 μl of PBS or DMSO containing test peptide (final concentration from 1 nM up to 100 μM).
2. Incubate the reaction mixtures for up to 48 h at room temperature. The equilibration time is DR-dependent and needs to be determined experimentally.
3. At the end of the incubation period apply all of the incubation mixtures on Bio-Spin columns (Bio-Rad) and centrifuge at 1100 *g* for 4 min.
4. Count excluded material (=complex of ^{125}I-labelled peptide with DR) that is collected into the collection tubes directly in a γ-scintillation counter.

3. Side chain scanning with p1-anchored designer peptide libraries

3.1 The prominent role of p1 anchors

The anchor closest to the N terminus of class II bound peptides is defined as the p1 anchor. The p1 anchor seems to be prominent, in that an Ala sub-

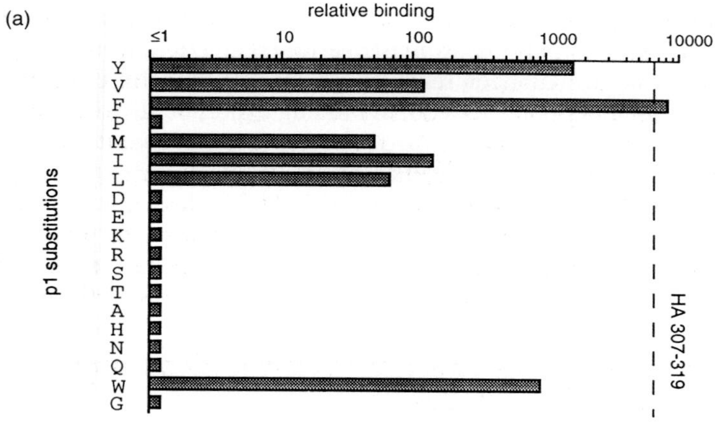

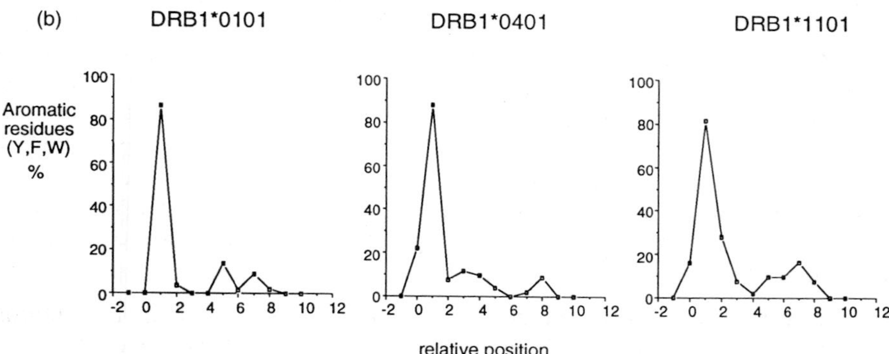

Figure 9. The prominent role of anchor position 1. (a) The p1 substitutions in the designer peptide YRSMAAAAA show the essential role of aromatic/aliphatic amino acid residues for binding to DRB1*0401. (b) The aligned sequence pools selected by three different HLA-DR alleles show a striking enrichment of aromatic residues near the NH_2 terminus of the peptide inserts.

stitution at this position abrogates peptide binding completely, whereas the elimination of other anchors results only in partial loss of binding affinity (9). The p1 anchor is conserved, since each of the DR-selected peptide pools show a striking enrichment of aromatic residues at this position (*Figure 9b*). *Figure 9a* illustrates that only the aromatic anchor residues Tyr, Phe, and Trp, that were identified by the M13 screening procedure, and to some extent the aliphatic residues Leu, Ile, Val, and Met, enable the designer peptides to bind with good affinity to DRB1*0401. These results fit well with the structure of

HLA-DR molecules, as revealed by X-ray crystallography (10) and computer modelling. The cleft contains only one deep pocket lined with non-polar residues capable of accommodating the dominant aromatic/aliphatic anchor residues. Three shallow pockets which may accommodate side chains of other anchor residues have also been described (10). The deep pocket is built by the invariant α-chain and by a fairly conserved part of the β-chain, justifying the conserved nature of the p1 anchor. Nevertheless, the composition of anchor residues in p1 is influenced by some allelic variations, such as the Gly/Val polymorphism at 86β, which changes the size of the p1 pocket, resulting in a preferential interaction with aromatic or aliphatic residues, respectively.

3.2 The principle of side chain scanning with p1-anchored designer peptide libraries

X-ray crystallographic studies indicate that different peptides bind with a similar conformation to HLA-DR molecules (10). This observation is supported by the analysis of the large number of peptide sequences selected from M13 phage display libraries. Although these peptides differ in their primary sequence, most of them share perfectly spaced anchor residues, suggesting an overall similar and sequence-independent peptide conformation. Experiments with designer peptides further support this hypothesis. Anchor and inhibitory residues lose their effect in different designer peptides if shifted by one position towards the N or C terminus. This indicates a low degree of flexibility in the peptide structure once bound to the MHC molecule. Crystallographic studies and peptide truncation experiments further indicate the importance of peptide main chain interaction for binding. Based on this information the principles for the side chain scanning with p1-anchored designer peptide libraries could be defined as follows:

(a) The obligatory p1 anchor together with the peptide main chain interactions determine the frame and the conformation of HLA-DR bound peptides.

(b) Side chain interactions have nearly no effect on the main chain conformation and consequently they are independent from each other. Therefore, we can assume that binding depends on the net result of anchor and inhibitory residues.

(c) The position of a particular side chain, with respect to the p1 anchor, will determine whether a given amino acid behaves as an anchor residue, as an inhibitory residue, or as a residue with neutral effect on binding.

To validate these concepts, positions p2 to p9 were scanned for the effect of each amino acid side chain on binding using high flux peptide binding assays and Ala-based designer peptide libraries. The examples in *Figure 10* demonstrate the power of this technology in detecting not only anchor residues previously identified by M13 libraries, but also position-specific inhibitory residues.

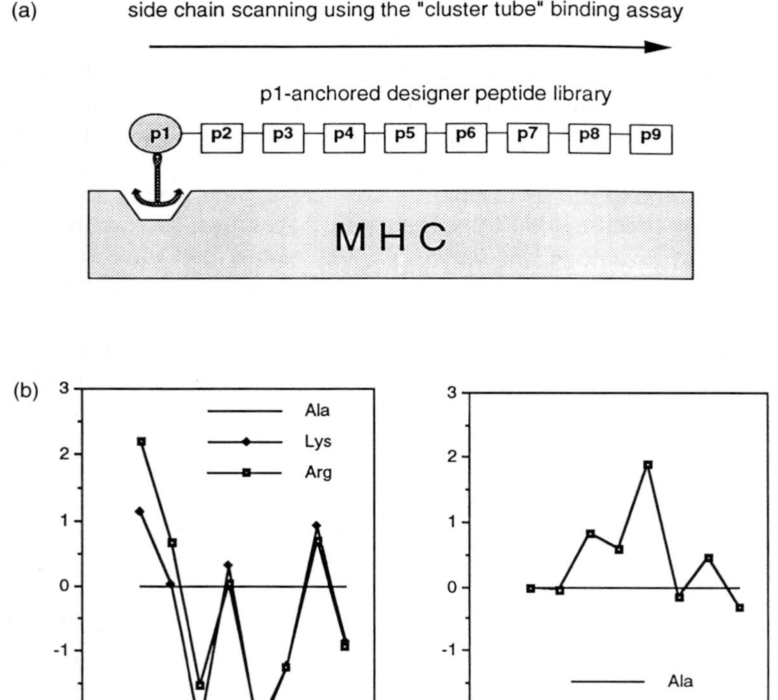

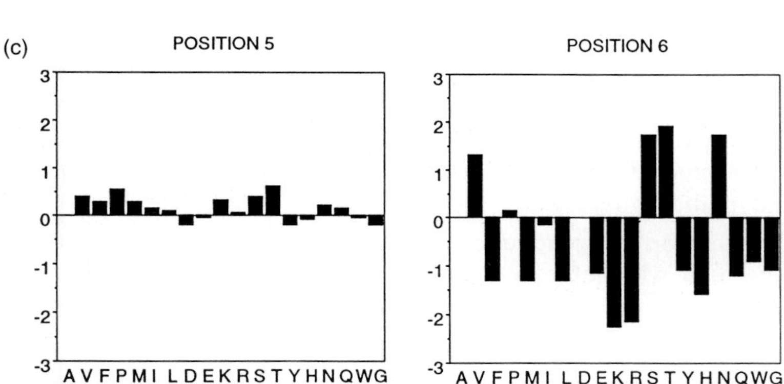

Figure 10. The principle of side chain scanning of p1-anchored designer peptide libraries. (a) Schema to demonstrate the anchoring of the peptide library. (b) Example 1; scanning of DRB1*0401 with Arg, Lys, and Thr, and identification of position-specific anchor and inhibitory residues. (c) Example 2: scanning with 19 amino acid residues at p5 and p6 on DRB1*0401. Only p6 is critical for the interaction with the MHC binding cleft.

3.3 The cluster tube peptide binding assay

A complete side chain scanning from p2 to p9 using a single p1-anchored designer peptide library includes 160 peptides. The 'cluster tube' peptide binding assay is a high flux competition assay which can process 160 peptides in two days (*Protocol 7*). This corresponds to 960 single assays using six dilutions for each peptide. The peptides are synthesized with multiple peptide synthesizers, such as the Advanced ChemTech 396 synthesizer. The purity of each peptide is determined by reverse-phase HPLC and the average purity should exceed 80%. Double or triple coupling procedures and at least a ten-fold surplus of Fmoc amino acids as compared to the free NH_2 termini of the resin are used to minimize the production of deletion mutants which would shift the frame of the peptides relative to the p1 anchor. Other impurities and slight variations in the amount of peptides are below the sensitivity of the cluster tube peptide binding assay and can be neglected.

The 'cluster tube' peptide binding assay is illustrated in *Figure 11*. First, peptides are simultaneously diluted in 96-well plates using a multichannel pipette. Next, the peptides are transferred to binding plates to which ^{125}I-labelled peptide and DR molecules are added. After incubation and addition of the biotinylated anti-DR antibody, the binding mixtures are transferred to 96 cluster tube racks, each tube containing streptavidin immobilized on agarose. The MHC/peptide/antibody complexes are bound to streptavidin, and the agarose in each tube is washed several times using a multichannel apparatus. The tubes are then counted directly in a γ-scintillation counter and the concentration of the designer peptides giving 50% inhibition of the ^{125}I-peptide/DR complex formation (IC_{50}) is determined.

Protocol 7. Cluster tube peptide binding assay

Equipment and reagents

- p1 anchor designer peptide library: all possible amino acid substitutions of a short Ala designer peptide with a p1 anchor at the NH_2 terminus
- Immunoaffinity purified DR antigen (see *Protocol 1*)
- 2 × binding buffer (see *Protocol 6*)
- Mix 1: ^{125}I-labelled DR binding peptides in PBS[a]
- Mix 2: DR antigen in 2 × binding buffer[a]
- Mix 3: biotinylated anti-DR antibody in PBS[a]
- BSA blocked streptavidin on 4% beaded agarose (see *Protocol 4*)
- Fluid delivering and removing devices (see *Figure 11b*)
- Cluster tubes (Costar) and 96-well plates (polypropylene)

Method

1. Dissolve peptides in DMSO to a final concentration of 2 mM and transfer 50 μl of each peptide into the (A) rows of 96-well plates.
2. With a 12-channel pipette add 50 μl of PBS into the (A) row wells and 180 μl of PBS:DMSO (1:1) into the (B)–(F) row wells of the 96-well

Protocol 7. *Continued*

dilution plates. Dilute simultaneously the peptides by serially transferring 20 μl from row (A) to row (F), so that each column corresponds to a series of dilution of a specific peptide.

3. Transfer with a 12-channel pipette 10 μl of each well from the dilution plates to the corresponding wells of the binding plates.
4. Add first 30 μl of mix 1 and then 30 μl of mix 2 to each well using multiple pipetting devices and incubate for at least 20 h at room temperature.
5. Add 30 μl of mix 3 to each well and incubate for another 3 h at room temperature.
6. Add 200 μl of streptavidin solid phase (total 20 μl settled volume) into each tube (row (A) to (F)) of the cluster tube racks.
7. Transfer with a 120 channel pipette 70 μl of each well from the binding plates to the corresponding tubes of the cluster tube racks and incubate for 10 min. Vortex the cluster tube racks several times during incubation.
8. Wash the streptavidin solid phase at least four times with PBS using specifically constructed devices (see *Figure 11b*) and transfer tubes to the γ-counter.

[a] The amount of ^{125}I-labelled peptide, DR antigen, and biotinylated antibody is determined empirically.

3.4 Optimization of designer peptide libraries

The affinity of the ^{125}I-labelled peptide in p1-anchored designer peptide libraries is crucial for the identification of side chain effects. The following formula is intended to be used as a guideline for the design of basic library peptides.

$$\Delta(1/IC_{50}) \approx \Delta\Sigma A + \Delta\Sigma I + \Delta L$$

The formula is based on experiments such as presented in *Figure 12* and can be explained as follows. The change (Δ) of the reverse IC_{50} of an HLA-DR binding peptide depends on the number and characteristics of anchor residues ($\Delta\Sigma A$), on the number and characteristics of inhibitory residues ($\Delta\Sigma I$), and on the peptide length (ΔL). For a given DR/^{125}I-peptide pair, the above formula is only applicable within a narrow range of values. For example, a basic peptide with high values of L and ΣA binds very tightly and consequently, the addition or removal of an anchor does not significantly change the reverse IC_{50}. Conversely, a peptide with very low values of L does not bind at all and the addition of an anchor is not likely to show any effect.

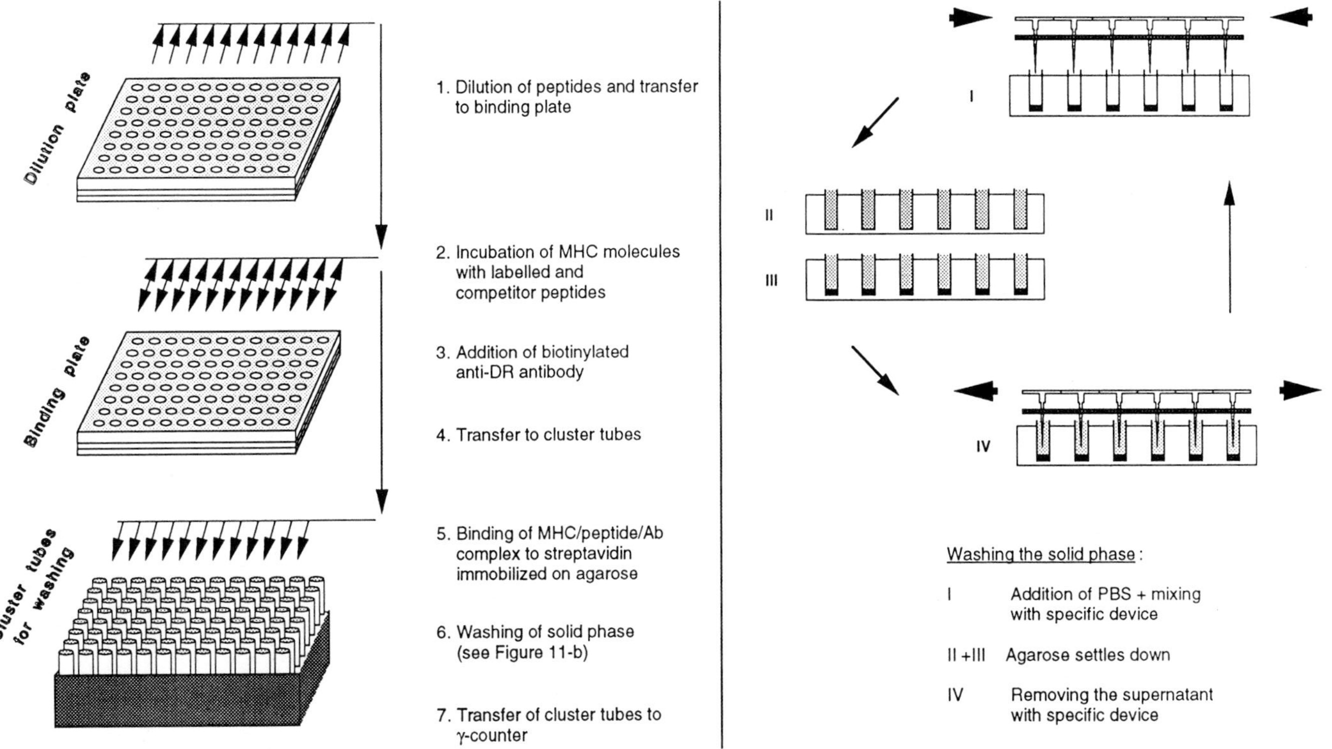

Figure 11. Schematic representative of the 'cluster tube' peptide binding assay.

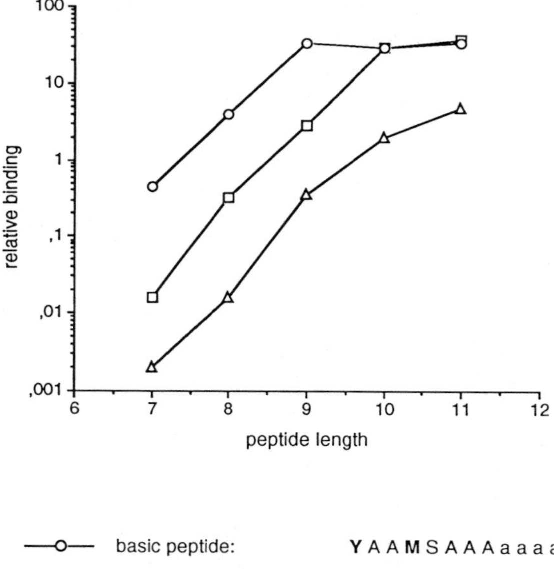

Figure 12. The number of anchor residues and the peptide length influence binding to the HLA-DRB1*0401 allele. Small letters indicate the truncated region of the peptides.

To increase the relative effect of $\Delta\Sigma A$ and $\Delta\Sigma I$, we suggest reducing the main chain interaction ($=L$) by using nonamers instead of better binding undecamers, e.g. FASAAAAAA. In case the resulting IC_{50} is still too high to allow the detection of negative effects, an additional anchor should be added, e.g. the conserved p2 Arg anchor: FRSAAAAAA. A second aromatic/aliphatic residue at p2 should be avoided to prevent a p1→p2 shift in the anchor frame. If both peptides give non-satisfying results we recommend optimization experiments as indicated in *Figure 12*.

The choice of the ^{125}I-labelled peptides is important for the side chain scanning. Ideally, they should have an affinity which is low enough to permit competition and, at the same time, high enough to be detectable as MHC/peptide complexes.

3.5 Analysis and interpretation of results

The IC_{50} data derived from side chain scanning of p1-anchored designer libraries are processed as follows: first, the IC_{50} data are normalized by dividing for the IC_{50} values obtained from the Ala-substitutions. This enables comparison of data among assays performed under slightly different conditions, e.g. assays carried out with different batches of DR proteins. Secondly, the reverse values of the normalized data are determined, and last, the logarithm of the reverse values is calculated. Zero indicates the Ala baseline, values below zero show inhibitory effects, and values above zero anchor effects. We consider only values equal or exceeding 1 or −1 as significant.

4. From side chain scanning to software: a quantitative prediction of class II/peptide binding

Methods to identify regions of protein sequences capable of binding MHC proteins would be very valuable for many immunological applications.

The binding data derived from side chain scanning of p1-anchored designer peptide libraries can be processed into a software predicting MHC binding regions in proteins. The software is based on the assumption that side chains interact with the MHC binding cleft independently from each other (see Section 3.2). Protein sequences are first scanned for p1 anchor residues. After having identified a p1 anchor, the program locates the amino acid residues at relative positions 2 to 9. Next, values obtained from the side chain scanning of p1-anchored peptide libraries (see Section 3.4) are assigned to each of the amino acid residues of the selected protein region. The sum of these values gives a score indicating the predicted peptide binding affinity (*Figure 13*).

The program listed in the Appendix is a simple prediction software, written with Microsoft QBASIC for Macintosh (11–13). *Protocol 8* explains how binding data obtained from peptide side chain scanning can be incorporated

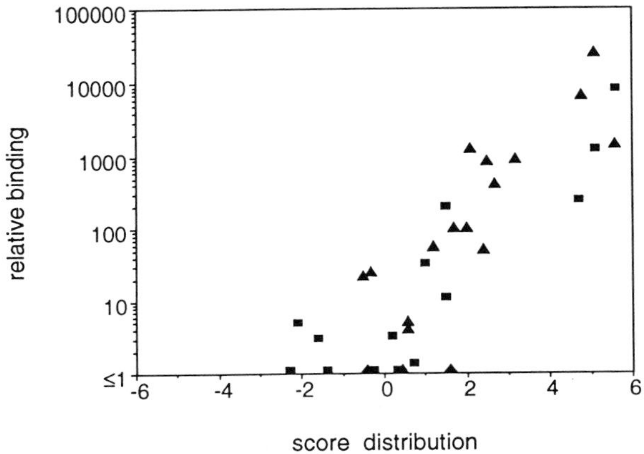

Figure 13. Correlation between peptide 'scores' and peptide binding affinity. The peptide scores of randomly selected peptides (13-mers) predicted to bind to DRB1*0401. (Prediction software, see Appendix.) The relative peptide binding is determined by the cluster tube peptide binding assay.

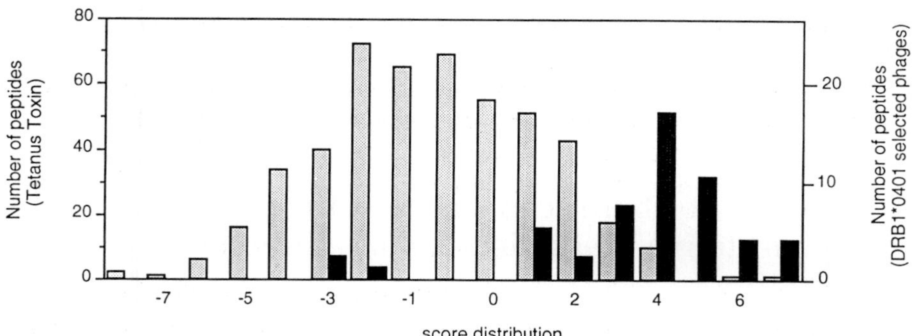

Figure 14. Peptide scores for M13 displayed peptides. The peptides of DRB1*0401-selected bacteriophage (black bars) show scores corresponding to high affinity binding (see *Figure 13*). Grey bars correspond to all peptides from the tetanus toxin sequence carrying a p1 anchor.

into the program. The program works with the Macintosh bottom and window routines and is therefore self-explaining once it is started.

As an example we have used this algorithm to analyse the set of peptides displayed by the DRB1*0401-selected phage (*Figure 14*). As expected most of the peptides score higher than 1.

Protocol 8. Transfer of binding data into software for the prediction of MHC binding peptides

1. Calculate the logarithm (log) of the reverse and normalized IC_{50} values derived from the side chain scanning of a p1-anchored designer peptide library (see Section 3.4).
2. Transfer values into the program listed in the Appendix, as outlined in the following example. If the amino acid residue Val has the value 2.1 at position 2, add the following line to the program:

   If p2$ = "V" then dr = dr + 2.1

   If the amino acid residue Asp has the negative value −1.3 at position 2, add this line:

   If p2$ = "D" then dr = dr − 1.3

   Add the values of position 2 for each amino acid residue to lines as shown in the examples. Use only the single letter code for the amino acid residues. Each line needs to end with "return". All lines are inserted between line 1010 and 2020 of the software (see Appendix).
3. Repeat the same for the positions 2 to 9. For position 3 change the variable to p3$, for position 4 to p4$ and so forth. The lines do not need to be numbered. The "if" and "then" change to the capital letters "IF"

and "THEN" after pressing "return", indicating a correct edition of the line.

4. The program does not distinguish between the anchor residues at position 1, meaning that each peptide with Tyr, Phe, Trp, Met, Ile, Leu, or Val is considered for calculation. If, for example, Tyr is not a p1 anchor as is the case in DR alleles with Val at 86β (see Section 3.1), add the following line:

If p1$ = "Y" then dr = dr − 10000

The low value −10000 will exclude any peptide starting with Tyr.

5. The p1 anchor position used in the basic peptide of the library attains the value zero, therefore there is no need to add a line to the software. If binding data are available demonstrating a difference of p1 anchors in binding, the log of the difference between the reverse IC_{50} should be determined and added as a line. For example, if peptides with Leu at p1 bind tenfold less than the basis peptide with Tyr at p1, the following line needs to be added:

If p1$ = "L" then dr = dr − 1

6. To screen a protein sequence for MHC binding peptides, either edit the amino acid sequence within the program or load it as textfile. Protein sequences from databases can be copied, pasted into any word processing software, and saved as textfile. Before the sequence is read by the program any 'spaces' or 'returns' should be removed from the sequence.

More recently, more complex prediction algorithms have been developed that enable the computational identification of (i) HLA class II ligands binding in a promiscuous or allele specific mode, and the identificatin of (ii) the effects of polymorphic residues on class II ligand specificity (14). For example, TEPITOPE is a recent Windows95 and WindowsNT application that permits the prediction and parallel display of ligands for twenty five HLA-DR alleles starting from any protein sequence (14).

Appendix

```
0010 REM******************** PREDICTION 3.0 ***********************
0020 REM *********************** input unit **********************
0030 einstein:TEXTFACE(1):TEXTSIZE(14):CLS:CLEAR:upperlevel = 1000
0040 WINDOW 2, "PREDICTION INPUT UNIT", (50,50)-(600,400),1
0050 LINE(10,10)-(520,10):LINE(10,250)-(520,250):LINE(10,10)-(10,250)
0060 LINE(520,10)-(520,250):upperlevel = 1000
0070 schleife:
0080    BUTTON 2,1,"EDIT SEQUENCE",(100,260)-(200,280)
0090    BUTTON 3,1,"LOAD SEQUENCE",(100,290)-(200,310)
0100    BUTTON 4,b1,"SAVE SEQUENCE",(100,320)-(200,340)
0110    BUTTON 5,b2,"<<< CALCULATE >>>",(210,320)-(520,340)
0120    BUTTON 6,b3,"START POSITION",(370,260)-(520,280)
0130    BUTTON 7,b4,"NAME OF PROTEIN",(210,290)-(360,310)
0140    BUTTON 8,1,"NEW",(20,285)-(70,315)
0150    BUTTON 9,0,"MHC CLASS II ALLELE",(370,290)-(520,310)
0160    BUTTON 10,b5,"SELECTION LEVEL",(210,260)-(360,280)
0170 WHILE DIALOG (O)<>1:WEND
0180 entscheidung0:
0190    IF DIALOG (1) <> 8 THEN GOTO entscheidung1
0200    GOTO einstein
0210 entscheidung1:
0220    IF DIALOG (1) <> 5 THEN GOTO entscheidung2
0230    sequence$ = EDIT$(1)
0240    WINDOW CLOSE 2:GOTO core
0250 entscheidung2:
0260    IF DIALOG (1) <> 2 THEN GOTO entscheidung3
0270    b1 = 1:sequence$=""
0280    BUTTON 4,b1,"SAVE SEQUENCE",(100,320)-(200,340)
0290    BUTTON 3,0"LOAD SEQUENCE",(100,290)-(200,310)
0300    BUTTON 2,0,EDIT SEQUENCE",(100,260)-(200,280)
0310    BUTTON 8,0,"NEW",(20,285)-(70,315)
0320    EDIT FIELD 1,"",(10,10)-(520,250)
0330    schleife1:
0340    act=DIALOG (O)
0350    IF act=1 THEN sequence$ = EDIT$(1) ELSE GOTO schleife1
0360    b5=1
0370 entscheidung3:
0380    IF DIALOG (1) <> THEN GOTO entscheidung4
0390    sequence$ = EDIT$(1)
0400    filename$ = FILES$(0)
0410    OPEN filename$ FOR OUTPUT AS #1
0420    PRINT#1,sequence$
0430    CLOSE#1
0440 entscheidung4:
0450    IF DIALOG (1) <>3 THEN GOTO entscheidung5
0460    filename$ = FILES$(1)
```

```
0470    IF filename$ = ""THEN GOTO schleife
0480    OPEN filename$ FOR INPUT AS #2
0490    WHILE (NOT EOF(2))
0500    LINE INPUT #2, sequence$
0510    EDIT FIELD 1,sequence$,(10,10)-(520,250)
0520    WEND
0530    CLOSE #2
0540    b5 =1:b1 = 1
0550    GOTO entscheidung5
0560 entscheidung5:
0570    IF DIALOG (1) <>7 THEN GOTO entscheidung6
0580    WINDOW 3,<<<<<<<< Please enter name of protein and press
return >>>>>>>>", (75,200)-(550,220),1
0590    INPUT protein$
0600    WINDOW CLOSE 3
0610 entscheidung6:
0620    IF DIALOG (1) <>6 THEN GOTO entscheidung8
0630    WINDOW 3, "Please enter start position of sequence and press
return", (75,200)-(550,220),1
0640    INPUT startposition:startposition = startposition - 1
0650    WINDOW CLOSE 3
0660 entscheidung8:
0670    IF DIALOG (1) <>10 THEN GOTO schleife
0680    b2 = 2:b3 = 1:b4 = 1
0690    WINDOW 3, "level of selection", (80,90)-(450,280),1
0700    BUTTON 58,1,"-- HIGH --", (30,40)-(300,60)
0710    BUTTON 59,1,"-- MEDIUM --", (30,70)-(300,90)
0720    BUTTON 60,1,"-- LOW --", (30,100)-(300,120)
0730    BUTTON 61,1,"-- NONE --", (30,130)-(300,150)
0740    BUTTON 62,1,"-- USER DEFINITION --", (30,160)-(300,180)
0750    WHILE DIALOG (0) <> 1:WEND
0760    IF DIALOG (1) = 58 THEN level = 4
0770    IF DIALOG (1) = 59 THEN level = 3
0780    IF DIALOG (1) = 60 THEN level = 2
0790    IF DIALOG (1) = 61 THEN level = -1000
0800    IF DIALOG 1 <> 62 THEN WINDOW CLOSE 3:GOTO schleife
0810    WINDOW CLOSE 3
0820    WINDOW 4, Please enter LOWER level and press return, (75, 200)-
        (550,220),1
0830    INPUT level
0840    WINDOW CLOSE 4
0850    WINDOW 5, Please enter UPPER level and press return", (75,200)-
(550,220),1
0860    INPUT upperlevel
0870    WINDOW CLOSE 4
0880    GOTO schleife
0890    REM ************** core program *****************************
0900 core:
```

```
0910    WINDOW 3, “<<<<<<<< YOUR MAC IS PROCESSING DATA >>>>>>>>≶,
        (100,200)-(500,200),1
0920    sequence$ = UCASE$(sequence$):laenge = LEN (sequence$)
0930    DIM binding peptide$ (3000):DIM aanfang (3000)
0940    DIM aaende (3000):DIM gruppe (3000)
0950    FOR ana = 1 TO (laenge − 8)
0960    p1$ = MID$(sequence$, ana, 1):p2$ = MID$(sequence$, (ana + 1), 1)
0970    p3$ = MID$(sequence$, (ana + 2), 1):o4$ = MID$(sequence$, (ana
        + 3), 1)
0980    p5$ = MID$(sequence$, (ana + 4), 1):p6$ = MID$(sequence$, (ana
        + 5), 1)
0990    p7$ = MID$(sequence$, (ana + 6), 1):p8$ = MID$(sequence$, (ana
        + 7), 1)
1000    p9$ = MID$(sequence$, (ana + 8), 1):dr = 0
1010    IF p1$ <> “Y” AND p1$ <> “W” AND p1$ <> “F” AND p1$ <> “V”
AND p1$ <> “M” AND p1$ <> “L” AND p1$ <> “I” THEN GOTO hell

PLACE BINDING DATA HERE AS DESCRIBED IN PROTOCOL 8

2020    peptidenumber = peptidenumber + 1
2030    IF dr < level THEN GOTO hell
2040    IF dr > upperlevel THEN GOTO hell
2050    bindingpeptide$ (peptidenumber) = MID$(sequence$, (ana), 9)
2060    aanfang (peptidenumber) = ana + startposition
2070    aaende (peptidenumber) = ana + 8 + startposition:gruppe
        (peptidenumber) = dr
2080    hell: NEXT ana
2090    REM ************************ results *******************
2100    WINDOW 3, RESULTS, (50,50)-(600,400),1
2110    leer$=" ":strich$=" - ":querstrich$="/":
2120    dpunkt$=":":bstrich$=" - ":level$="level"
2130 FOR r = 1 TO laenge
2140    IF bindingpeptide$ (r) =""THEN GOTO schleifer
2150    nichts = 1
2160    PRINT protein$;:PRINT aanfang (r);:PRINT bstrich$;
2170    PRINT aaende (r);:PRINT dpunkt$;
2180    PRINT bindingpeptide$ (r);:PRINT querstrich$;:PRINT
level$;:PRINT gruppe (r)
2190    schleifer:
2200 NEXT r
2210    IF nichts = 0 THEN PRINT “NO PEPTIDES FOUND FOR SELECTED LEVEL”
2220    savebutton = 1
2230 schleifli:
2240    BUTTON 100,1,"click to end program",(200,320)-(350,340)
2250    BUTTON 101,savebutton,"click to save results",(370,320)-
(520,340)
2260    BUTTON 102,1,"new prediction",(30,320)-(180,340)
2270    WHILE DIALOG (O) <> 1:WEND
```

```
2280 entscheidung100:
2290    IF DIALOG (1) = 100 THEN END
2300 entscheidung102:
2310    IF DIALOG (1) = 102 THEN GOTO einstein
2320 entscheidung101:
2330    IF DIALOG (1) <> 101 THEN GOTO schleifli
2340    savefile$ = FILES$(0)
2350    IF savefile$ = ""THEN GOTO schleifli
2360    OPEN savefile$ FOR OUTPUT AS #2
2370    FOR r2 = 1 TO laenge
2380 IF bindingpeptide$ (r2) = ""THEN GOTO schleifer2
2390    PRINT #2, protein$;:PRINT #2, aanfang (r2);
2400    PRINT #2, bstrich$;:PRINT #2, aaende (r2);
2410    PRINT #2, punkt$;:PRINT #2, bindingpeptide$ (r2);:PRINT #2,
querstrich$;
2420    PRINT #2, level$;:PRINT #2, gruppe (r2);:PRINT #2,
allele$;:PRINT #2, leer$
2430    schleifer2:
2440    NEXT r2
2450    CLOSE #2:WINDOW 3, "RESULTS", (50,50)-(600,400),1:savebutton =
0
2460    GOTO schleifli
2470    REM ************************ list end ******************
```


Acknowledgements

We thank Luciano Adorini and Elisa Bono for critically reading the manuscript.

References

1. Sinigaglia, F. and Hammer, J. (1994). *Curr. Opin. Immunol.*, **6**, 52.
2. Parmeley, S. F. and Smith, G. P. (1988). *Gene*, **73**, 305.
3. Scott, J. K. and Smith, G. P. (1990). *Science*, **249**, 386.
4. Devlin, J. J., Panganiban, L. C., and Devlin, P. E. (1990). *Science*, **249**, 404.
5. Cwirla, S. E., Peters, E. A., Barrett, R. W., and Dover, W. J. (1990). *Proc. Natl. Acad. Sci. USA*, **87**, 6378.
6. Hammer, J., Takacs, B., and Sinigaglia, F. (1992). *J. Exp. Med.*, **176**, 1007.
7. Sambrook, J., Fritsch, E. F., and Maniatis, T. (ed.) (1989). *Molecular cloning, a laboratory manual.* Cold Spring Harbor Press, Cold Spring Harbor, NY.
8. Hammer, J., Valsasnini, P., Tolba, K., Bolin, D., Higelin, J., Takacs, B., *et al.* (1993). *Cell*, **74**, 197.
9. Hammer, J., Belunis, C., Bolin, D., Papadopulos, J., Walsky, R., Higelin, J., *et al.* (1994). *Proc. Natl. Acad. Sci. USA.*, **91**, 4456
10. Brown, J. H., Jardetzky, T. S., Gorga, J. C., Stern, L. J., Urban, R. G., Strominger, J. L., *et al.* (1993). *Nature*, **364**, 33.

11. Hammer, J., Bono, E., Gallazzi, F., Belunis, C., Nagy, Z., and F. Sinigaglia (1994). Precise prediction of major histocompatibility complex class II peptide interaction based on peptide side chain scanning. *J Exp Med* **180**, 2353.
12. Hammer, J., Gallazzi, F., Bono, E., Karr, R. W., Guenot, J., Valsasnini, P., Nagy, Z. A., and F. Sinigaglia (1995). Peptide binding specificity of HLA-DR4 molecules: correlation with rheumatoid arthritis association. *J Exp Med* **181**, 1847.
13. Hammer, J. (1995) New methods to predict MHC-binding sequences within protein antigens. *Current Opinion in Immunology* **7**, 263.
14. Hammer, J., Sturniolo, T., Sinigaglia, F. (1997). HLA class II binding specificity and autoimmunity. *Advances in Immunology* **66**.

A1

List of suppliers

Amersham

Amersham International plc., Lincoln Place, Green End, Aylesbury, Buckinghamshire HP20 2TP, UK.

Amersham Corporation, 2636 South Clearbrook Drive, Arlington Heights, IL 60005, USA.

Anderman

Anderman and Co. Ltd., 145 London Road, Kingston-Upon-Thames, Surrey KT17 7NH, UK.

Beckman Instruments

Beckman Instruments UK Ltd., Progress Road, Sands Industrial Estate, High Wycombe, Bucks HP12 4JL, UK.

Beckman Instruments Inc., PO Box 3100, 2500 Harbor Boulevard, Fullerton, CA 92634, USA.

Becton Dickinson

Becton Dickinson and Co., Between Towns Road, Cowley, Oxford OX4 3LY, UK.

Becton Dickinson and Co., 2 Bridgewater Lane, Lincoln Park, NJ 07035, USA.

Bio

Bio 101 Inc., c/o Statech Scientific Ltd, 61–63 Dudley Street, Luton, Bedfordshire LU2 0HP, UK.

Bio 101 Inc., PO Box 2284, La Jolla, CA 92038–2284, USA.

Bio-Rad Laboratories

Bio-Rad Laboratories Ltd., Bio-Rad House, Maylands Avenue, Hemel Hempstead HP2 7TD, UK.

Bio-Rad Laboratories, Division Headquarters, 3300 Regatta Boulevard, Richmond, CA 94804, USA.

Boehringer Mannheim

Boehringer Mannheim UK (Diagnostics and Biochemicals) Ltd, Bell Lane, Lewes, East Sussex BN17 1LG, UK.

Boehringer Mannheim Corporation, Biochemical Products, 9115 Hague Road, P.O. Box 504 Indianapolis, IN 46250–0414, USA.

Boehringer Mannheim Biochemica, GmbH, Sandhofer Str. 116, Postfach 310120 D-6800 Ma 31, Germany.

British Drug Houses (BDH) Ltd, Poole, Dorset, UK.

Difco Laboratories

Difco Laboratories Ltd., P.O. Box 14B, Central Avenue, West Molesey, Surrey KT8 2SE, UK.

Difco Laboratories, P.O. Box 331058, Detroit, MI 48232–7058, USA.

Du Pont

Dupont (UK) Ltd., Industrial Products Division, Wedgwood Way, Stevenage, Herts, SG1 4Q, UK.

Du Pont Co. (Biotechnology Systems Division), P.O. Box 80024, Wilmington, DE 19880–002, USA.

European Collection of Animal Cell Culture, Division of Biologics, PHLS Centre for Applied Microbiology and Research, Porton Down, Salisbury, Wilts SP4 0JG, UK.

Falcon (Falcon is a registered trademark of Becton Dickinson and Co.).

Fisher Scientific Co., 711 Forbest Avenue, Pittsburgh, PA 15219–4785, USA.

Flow Laboratories, Woodcock Hill, Harefield Road, Rickmansworth, Herts. WD3 1PQ, UK.

Fluka

Fluka-Chemie AG, CH-9470, Buchs, Switzerland.

Fluka Chemicals Ltd., The Old Brickyard, New Road, Gillingham, Dorset SP8 4JL, UK.

Gibco BRL

Gibco BRL (Life Technologies Ltd.), Trident House, Renfrew Road, Paisley PA3 4EF, UK.

Gibco BRL (Life Technologies Inc.), 3175 Staler Road, Grand Island, NY 14072–0068, USA.

Arnold R. Horwell, 73 Maygrove Road, West Hampstead, London NW6 2BP, UK.

Hybaid

Hybaid Ltd., 111–113 Waldegrave Road, Teddington, Middlesex TW11 8LL, UK.

Hybaid, National Labnet Corporation, P.O. Box 841, Woodbridge, NJ. 07095, USA.

HyClone Laboratories 1725 South HyClone Road, Logan, UT 84321, USA.

International Biotechnologies Inc., 25 Science Park, New Haven, Connecticut 06535, USA.

Invitrogen Corporation

Invitrogen Corporation 3985 B Sorrenton Valley Building, San Diego, CA. 92121, USA.

Invitrogen Corporation c/o British Biotechnology Products Ltd., 4–10 The Quadrant, Barton Lane, Abingdon, OX14 3YS, UK.

Kodak: Eastman Fine Chemicals 343 State Street, Rochester, NY, USA.

Life Technologies Inc., 8451 Helgerman Court, Gaithersburg, MN 20877, USA.

Merck

Merck Industries Inc., 5 Skyline Drive, Nawthorne, NY 10532, USA.

Merck, Frankfurter Strasse, 250, Postfach 4119, D-64293, Germany.

Millipore

Millipore (UK) Ltd., The Boulevard, Blackmoor Lane, Watford, Herts WD1 8YW, UK.

Millipore Corp./Biosearch, P.O. Box 255, 80 Ashby Road, Bedford, MA 01730, USA.

New England Biolabs (NBL)

New England Biolabs (NBL), 32 Tozer Road, Beverley, MA 01915–5510, USA.

New England Biolabs (NBL), c/o CP Labs Ltd., P.O. Box 22, Bishops Stortford, Herts CM23 3DH, UK.

Nikon Corporation, Fuji Building, 2–3 Marunouchi 3-chome, Chiyoda-ku, Tokyo, Japan.

Perkin-Elmer

Perkin-Elmer Ltd., Maxwell Road, Beaconsfield, Bucks. HP9 1QA, UK.

Perkin-Elmer Ltd., Post Office Lane, Beaconsfield, Bucks, HP9 1QA, UK.

Perkin-Elmer-Cetus (The Perkin-Elmer Corporation), 761 Main Avenue, Norwalk, CT 0689, USA.

Pharmacia Biotech Europe Procordia EuroCentre, Rue de la Fuse-e 62, B-1130 Brussels, Belgium.

Pharmacia Biosystems

Pharmacia Biosystems Ltd. (Biotechnology Division), Davy Avenue, Knowlhill, Milton Keynes MK5 8PH, UK.

Pharmacia LKB Biotechnology AB, Björngatan 30, S-75182 Uppsala, Sweden.

Promega

Promega Ltd., Delta House, Enterprise Road, Chilworth Research Centre, Southampton, UK.

Promega Corporation, 2800 Woods Hollow Road, Madison, WI 53711–5399, USA.

Qiagen

Qiagen Inc., c/o Hybaid, 111–113 Waldegrave Road, Teddington, Middlesex, TW11 8LL, UK.

Qiagen Inc., 9259 Eton Avenue, Chatsworth, CA 91311, USA.

Schleicher and Schuell

Schleicher and Schuell Inc., Keene, NH 03431A, USA.

Schleicher and Schuell Inc., D-3354 Dassel, Germany. Schleicher and Schuell Inc., c/o Andermann and Company Ltd.

Shandon Scientific Ltd., Chadwick Road, Astmoor, Runcorn, Cheshire WA7 1PR, UK.

Sigma Chemical Company

Sigma Chemical Company (UK), Fancy Road, Poole, Dorset BH17 7NH, UK.

Sigma Chemical Company, 3050 Spruce Street, P.O. Box 14508, St. Louis, MO 63178–9916.

Sorvall DuPont Company, Biotechnology Division, P.O. Box 80022, Wilmington, DE 19880–0022, USA.

Stratagene

Stratagene Ltd., Unit 140, Cambridge Innovation Centre, Milton Road, Cambridge CB4 4FG, UK.

Strategene Inc., 11011 North Torrey Pines Road, La Jolla, CA 92037, USA.

United States Biochemical, P.O. Box 22400, Cleveland, OH 44122, USA.

Wellcome Reagents, Langley Court, Beckenham, Kent BR3 3BS, UK.

Index